Applied Civil Engineering Risk Analysis

Robb Eric S. Moss, Ph.D. P.E. F.ASCE

ISBN-10: 0989889602

ISBN-13: 978-0-9898896-0-5

CONTENTS

FORWARD

This text came out of my course notes from a new course that I developed when starting at Cal Poly as an assistant professor in 2006. I saw a need to reach a broad swath of upper division undergraduate and lower division graduate students with the concepts of risk analysis within a Civil Engineering context. The objective of this text is to introduce fundamental applied probability and reliability methods within one term of study so that, as the course outcome, students can perform first-order probability of failure analysis on Civil Engineering problems. To achieve this outcome I present a simplified or streamlined approach to the material:

- Only basic necessary statistical methods are presented, focusing on an intuitive understanding of the data and what it can tell us about past events.
- Probability is presented in a systematic manner to make solving the often tricky probability problems more tractable.
- Only the normal and lognormal distributions are used in this text because of their useful mathematical properties and because they can often approximate a solution adequately for a first-order analysis.
- Functions of random variables are presented in a canonical manner to make solving straight forward.
- A comprehensive set of reliability methods is presented, with emphasis on methods that can be solved with minimal background in higher math.
- The tools of decision analysis, systems analysis, and other related topics are presented in an applied and usable manner.
- Many examples covering a wide range of problems are presented.

My hope for this text is that; 1) readers are left with an appreciation of the uncertainty that exists in Civil Engineering problems and are able to quantify and treat the uncertainty in a proper way to enhance civil engineering design, and 2) that readers can tackle any problem and arrive at a first-order estimate of the probability of failure which will enable them to determine if further effort is necessary in refining the analysis.

Robb Moss, Ph.D. P.E. F.ASCE

ACKNOWLEDGMENTS

The intellectual support for this text has been provided by those whose shoulders I stand upon; those who paved the way in applied probability, civil engineering statistics, and structural reliability. Armen Der Kiureghian introduced me to and guided me through the concepts and methods in reliability, and set me on a Bayesian path. Ray Seed provided the initial opportunity and support for diving into this field. Alfredo Ang and Wilson Tang provided the foundation textbooks in probability and applied statistics with their 1st editions of Volume 1 and (the hard to find) Volume 2, as well as the excellent 2nd edition of Volume 1 that I used in teaching parts of this course. Fred Kulhawy and Paul Mayne with their work in defining uncertainty in geotechnical engineering in its various forms opened my eyes to aleatory and epistemic uncertainty. The seminal textbook by Jack Benjamin and Allin Cornell in which they canonized much of this material from its nascent form, as well as the accessible textbook by Milton Harr which included reliability, both provided a strong background of material to draw from. Gregory Baecher and John Christian and their comprehensive text on uncertainty in geotechnical engineering delved into the specifics of my chosen field and led to my better understanding of the material.

The logistical support for this book has been provided by Cal Poly in the form of a sabbatical to pursue this writing project. Other logistical support came from Montana State University and the Universidad de Concepcion for providing a good place to "hide out" as a visiting professor.

Emotional support for this endeavor has been provided by my family (Aleksandra, Konrad, and Sebastian), my colleagues, my students (Justin Hollenback in particular), and my friends …

Thanks to you all!

1 INTRODUCTION

Uncertainty exists. It exists in civil engineering problems, both in the engineering analysis and the engineering design. The key to dealing with uncertainty is properly quantifying it and then addressing it in a systematic manner. Some examples of uncertainty that exist in Civil Engineering* are presented in the following chapter, but first we need to define uncertainty which comes in two forms;

1. Aleatory, and
2. Epistemic.

Aleatory uncertainty, with the root "alea" derived from the Latin for rolling of dice, is the inherent or natural randomness. Whereas, epistemic uncertainty, stems from the lack of data, measurement error, improper mathematical modeling, or missing explanatory variables. Aleatory is the irreducible uncertainty in a particular problem. Epistemic is the reducible uncertainty that can be minimized through more study, additional data collection, better modeling, and other steps to better quantify the problem; it is the uncertainty that we as observers introduce into the problem. These two forms of uncertainty are often difficult if not impossible to separate, and can be hard to define depending on the problem and the perspective. Nonetheless the concepts of aleatory and epistemic uncertainty hold true.

To help clarify what may be considered aleatory and what may be considered epistemic, let us evaluate the breaking strength of concrete cylinders in a compression test. If we had single batch of concrete cylinders which we tested using the exact same loading equipment, the variability of the breaking strength values could be considered aleatory, but if there was any bias or imprecision due to the load measuring equipment (e.g., load cell out of calibration) that would be considered epistemic. If a second batch was tested and showed on average markedly different breaking strength, then the manufacturing process could be the culprit. If tighter control could be exerted on the manufacturing process to reduce the variability between batches then this batch variability would be considered epistemic. And if all

* Throughout this text the reference to Civil Engineering is shorthand for referring to Civil (Transportation, Structural, Geotechnical, Water) and related fields such as Environmental Engineering, Construction Engineering, Engineering Geology, etc. The methods presented in this text can apply to all related fields equally.

the breaking strength results were averaged, and the average value was used in an engineering calculation without including the variability, this would introduce additional epistemic uncertainty into the engineering process.

In most Civil Engineering problems we start out with some data on the system we are to analyze, and in almost all Civil Engineering problems we can frame the problem as one of load vs. resistance. Load is what is being required of the system. Resistance is what the system can withstand. [Load is synonymous with demand or stress, resistance is synonymous with strength or capacity]. We are interested in how reliable the system is, or conversely what is the probability of failure (P_f) in situations we would like to avoid such as; bridge collapse, slope failure, grid lock, culvert failure, chemical overload, project incompletion…

The following figures provide examples of the uncertainty that exists within Civil Engineering.

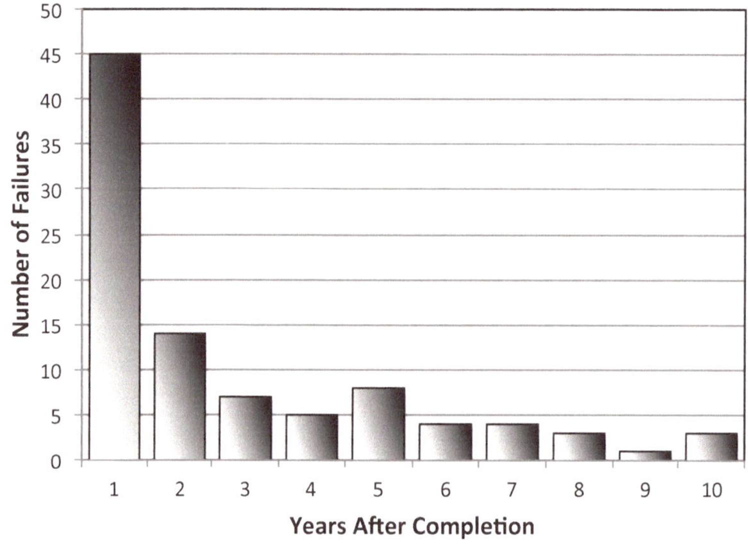

Figure 1.1 Statistics of dam failure in the United States (after van Gelder, 2000, Committee on Hydraulic and Geotechnical Engineering, Delft University of Technology).

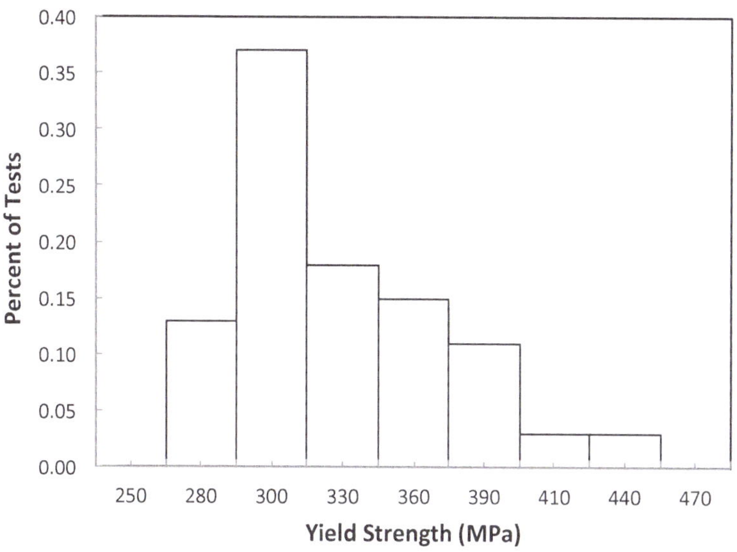

Figure 1.2 Yield strength of steel reinforcing bars (after Julian, 1957, Journal of Structural Division, ASCE, Vol.83).

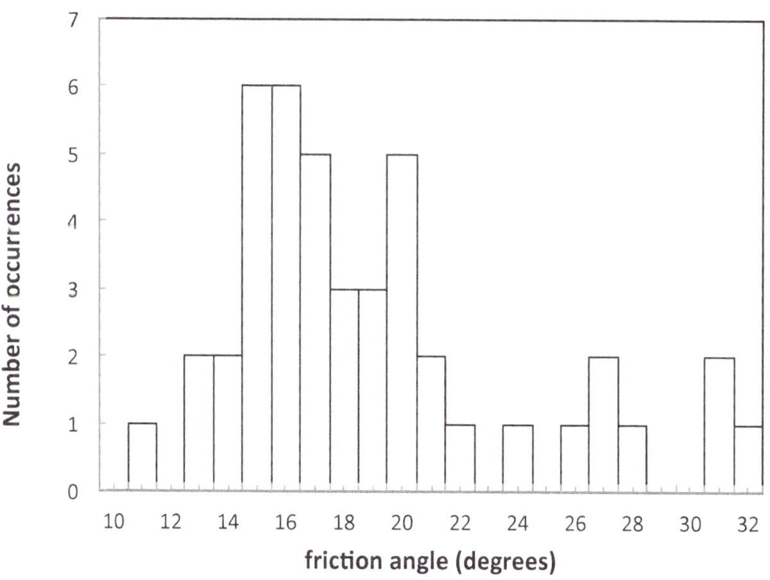

Figure 1.3 Residual friction angle of mudstone (after Becker et al., 1998, Canadian Geotechnical Journal, Vol.35).

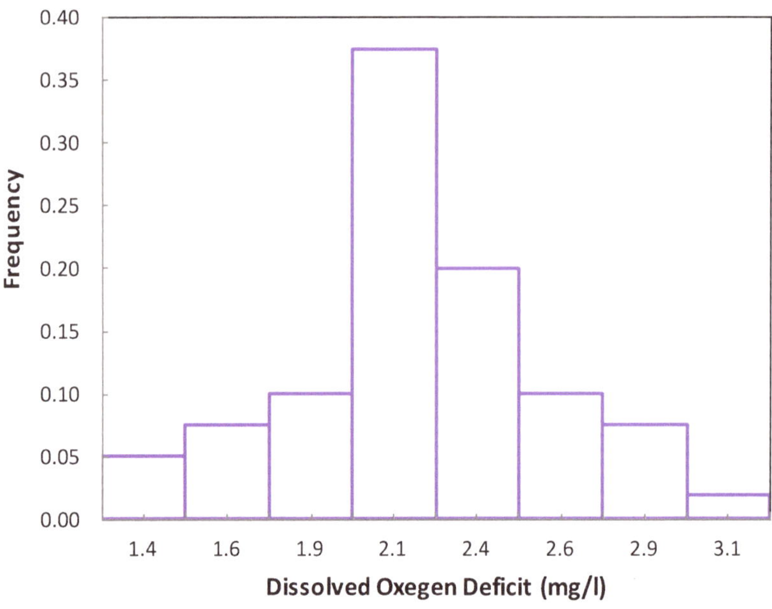

Figure 1.4 DO deficit in a river (after Kothandaraman and Ewing, 1969, Journal of Water Resources Control Federation, Feb).

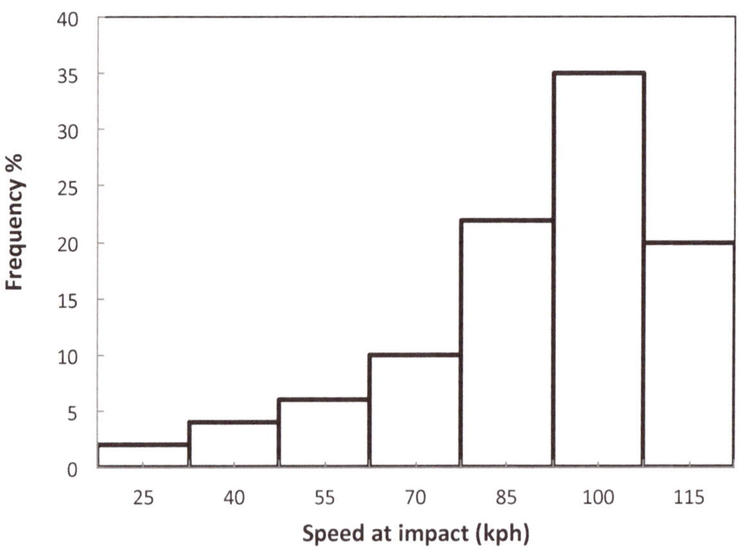

Figure 1.5 Vehicle impact speed of passenger car accidents (after Viner, 1972, Journal of Transportation Engineering, ASCE, Vol.98).

Example: Cantilever Beam

To demonstrate uncertainty in an engineering problem we will look at the deflection of a simple cantilever beam. The deflection at the end of the beam due to a point load is described by:

$$\Delta = \frac{PL^3}{3EI}$$

Here we will treat the load (P) and the length (L) of the beam as deterministic, meaning there is no uncertainty in the values. But if the stiffness (EI) of the beam is uncertain, due to manufacturing and material variability, then the deflection would also vary.

$P = 10kN$
$L = 1m$

$mean\ EI = 80\ kN \cdot m^2$
$std\quad EI = 20\ kN \cdot m^2$

The *mean* stands for the typical or average value and *std* for the standard deviation defining the uncertainty (these will be formally defined in subsequent chapters). If we do a simple sensitivity study by including the uncertainty:

$$mean \qquad \Delta = \frac{10\ kN \cdot m^3}{240\ kN \cdot m^2} = 0.042\ m$$

$$mean + std \quad \Delta = \frac{10\ kN \cdot m^3}{300\ kN \cdot m^2} = 0.033\ m$$

$$mean - std \quad \Delta = \frac{10\ kN \cdot m^3}{180\ kN \cdot m^2} = 0.056\ m$$

We can see that the variability in material properties can have an impact on the deflections. If the structural tolerance for deflection is < 0.045 m then there is a probability of unacceptable deformations and/or failure.

The bulk of this text is focused on estimating the influence of uncertainty on the probability of failure by;

- performing statistics on the data,
- applying the rules of probability,
- and defining failure within a reliability framework to calculate the probability of failure.

Statistics is simply a way of evaluating data. It is a means of interrogating past events. Probability is a method of projecting trends, or forecasting the future. Failure is defined subjectively by some engineering criteria; displacement, cracking, collapse, phase transformation, uptake, crossing...

Failure has traditionally been defined in Civil Engineering in an ASD/WSD (Allowable Stress Design/Working Stress Design) formulation using a factor of safety.

$$Factor\ of\ Safety = Strength/Stress \qquad equation\ 1.1$$

A safe design is when the Factor of Saftey ≥ 1. Here the uncertainty is often not quantified, and when it is quantified will often be lumped into the resistance side of the problem. As will be shown later in this text a more accurate means of defining failure is by using the limit state formulation or margin of safety.

$$Margin\ of\ Safety = Strength - Stress \qquad equation\ 1.2$$

A safe design is when Margin of Safety ≥ 0. Here the uncertainty is often quantified, and separated into uncertainty contributed from the load and from the resistance, respectively.

The probability of failure is then how near to the failure criteria a particular problem lies. Uncertainty is an inherent part of solving for the probability of failure. Without measuring the uncertainty the probability of failure is meaningless. Probability comes in various forms and the philosophical underpinnings of probability are quite complex and are open to many interpretations. Many authors have delved into the meaning of probability (e.g., Laplace, 1814) and readers are encouraged to follow up on this if there is interest, but for our purposes we assume that probability exists as a tool for forecasting future events or trends. Here we will define two camps with respect to probability, the;

- frequentist approach (or classic approach), and
- degree-of-belief approach (or Bayesian approach).

The frequentist approach relies on a large number of samples to define probability. Frequentist probability works well for, say, Manufacturing Engineering where man-made objects are produced in large numbers and the probability of a faulty item can be forecast based on a large database of objects tested to failure.

The degree-of-belief approach will result in the same probability if enough data exists for a particular problem but deals with situations where data is scarce, is to expensive or infeasible to collect, and a probability must be arrived at to move forward with the analysis. The types of problems that this approach lends to are; the probability of flooding at a particular site, the probability of liquefaction given the soil conditions, the probability of a column buckling under a load, or the probability of a contaminant moving into an aquifer... This form or probability is often necessary in Civil Engineering because engineered features tend to be unique both in the resistance and the load and there may be little or no data to draw from.

In practice the dual nature of probability can aid in achieving an engineering solution. One approach does not negate the other, and both approaches are equally valid and can be proven with mathematical rigor (de Finetti, 1972, 1974).

Once we have determined the probability of failure we can frame the problem with risk, where risk is defined in engineering as:

$$Risk = (Probability\ of\ Failure) \cdot (Consequences) \qquad equation\ 1.3$$

We move through the steps of analyzing the data, assessing the probability, defining failure, estimating the probability of failure, and measuring the consequences. We can then state what the risk is, usually quoted as an order of magnitude within some time frame. The consequences of failure is a scaling metric, often a monetary value or life loss number, used to rank the potential failure. The risk value can then be used to compare with other potential failures or types of loss to achieve a comprehensive balanced assessment of all failure modes and their consequences; a level assessment for making rational decisions.

For example, if we take the annual probability of levee failure in a specific location due to seismic loading to be on the order of 10^{-6}, this means there is a 1/1,000,000 chance of levee failure in any given year due to seismic loading. Let's say that there is an urban area behind the levee, and if levee breach and flooding were to occur, the loss of life is estimated to be on average 10 people per failure. The annual risk of life loss due to seismic failure of a levee is then 10^{-5}. Compare this to the annual risk of dying due to an auto accident of $6.0(10^{-4})$ which is in this example 60 times higher.

There are two distinct forms of risk; voluntary and involuntary. Voluntary risk is something that we subject ourselves to consciously, involuntary risk is when we are subjected to the risk without complete knowledge. The boundary is quite fuzzy and hard to define, and migrates depending on several factors. For instance, the annual risk of dying in an commercial airplane is less than 10^{-4} which has been held stable for years. In the 70's there were a spate of plane accidents which pushed the risk higher than 10^{-4}. This resulted in public outcry, legislation requiring stricter standards and safety oversight, and the decrease of the risk back down below the 10^{-4} range. Even though we choose to fly, the risk of dying in a commercial flight is considered by society to be involuntary and a 1/10,000 chance of death is what society deems as acceptable.

Contrast this with types of voluntary annualized risk of life loss that we subject ourselves to regularly which fall in the 1/100 and 1/1,000 range; risk 100 to 1000 times greater than involuntary (Table 1). It's also important to note that perceived risk is not equal to real risk, as people are in general not good at perceiving how "risky" things are. People tend to be afraid of dying via shark or bear attacks, but it is roughly 10,000 times higher annual odds that you will die in an accident in your home (e.g., falling from a ladder) than being attacked by a large predator.

Table 1.1: Risk Comparison

Type of Risk	Annual Risk of Death	Annual Odds of Death
Smoking a pack a day[1]	$3.6(10^{-3})$	1 in 278
Himalayan mountaineering[1]	$2.8(10^{-3})$	1 in 356
Cancer (all types)[6]	$2.6(10^{-3})$	1 in 387
Motor vehicle accidents[3]	$6.0(10^{-4})$	1 in 1,667
Killed in any war[2]	$2.4(10^{-4})$	1 in 4,167
Homocide[4]	$1.8(10^{-4})$	1 in 5,556
Air travel[3]	$7.0(10^{-5})$	1 in 14,286
Flood[5]	$1.0(10^{-5})$	1 in 100,000
Tornadoes[3]	$4.0(10^{-6})$	1 in 250,000
Hurricanes[3]	$4.0(10^{-7})$	1 in 2,500,000
[1]Wilson and Crouch (1987), [2]Small and Singer (1982), [3]US Nuclear Regulatory Commission (1975), [4]Holmes and De Burger (1988), [5]Lichtenstein et al. (1982), [6]http://www.hse.gov.uk/education/statistics.htm		

Chapter Summary

- Uncertainty exists in Civil Engineering analysis and design.
- Uncertainty comes in two forms; **aleatory** and **epistemic**.
- Almost all problems in Civil Engineering can be framed as **load** vs. **resistance** (synonymous with stress vs. strength or demand vs. capacity).
- Determining the **probability of failure** requires statistics to interrogate the data, probability to forecast future trends, and a definition of failure within a reliability framework.
- Probability can be viewed from a frequentist or degree-of-belief approach. The material in this text leans towards the degree-of-belief or Bayesian approach.
- **Risk** is the product of the probability of failure and the consequences, usually reported as an order of magnitude (e.g., annual risk of life loss on the order of 10^{-4}).
- Risk can be categorized as voluntary or involuntary depending on how it's applied and perceived.

2 DATA ANALYSIS / SAMPLE STATISTICS

Statistics is a means of interrogating past events. If we have some data on previous events, statistics are a set of tools that allow us to quantify the trends of these past events. Data in Civil Engineering are often scarce due to the unique nature of the engineered features (e.g., bridges, buildings, levees, tunnels, pipelines), but there are situations where we can assume that the data is similar enough in its variability, which is termed homoscedastic, to represent the same type of engineering problem. In these cases we need to estimate the trends of an occurrence of some event related to the engineered feature. How do we quantify uncertainty given some observations? The methods covered in this chapter include;

- Frequency plots (aka histograms),
- Sample statistics (for single and multiple variables), and
- Estimation methods.

A variable is defined as something that can assume a mathematical value for engineering calculation purposes†. Take for instance the equation of a line, $Y = mX + b$. Assuming that we know the slope, *m*, and the intercept, *b*, then *X* is a variable that can assume a range of mathematical values to calculate *Y*.

Histograms

A histogram is a means of plotting up data on a single variable to observe the frequency of occurrence. It is a bar chart showing how often the measured event occurs. Here we will take some annual rainfall data and look at what a histogram can tell us about rainfall events. Rainfall data was collected for 49

† A variable will be denoted in this text by a capital letter, *X*, whereas a specific numerical value of that variable will be denoted by a lower case letter, *x*. When a variable assumes a specific value we can describe it as *X*=*x*.

consecutive years and reported as the annual accumulation (usually in centimeters)‡ as shown in 2.1.

Table 2.1: Annual rainfall data for 49 consecutive years.

18	16	18	26	24	18
24	22	24	20	22	20
22	20	16	26	18	22
20	18	24	20	18	26
16	24	20	16	20	20
22	20	18	18	22	22
26	22	22	18	20	20
24	26	20	16	22	24
18					

The frequency or count for a particular annual rainfall value is then plotted as a histogram (Figure 2.1). If we label the rainfall data with the variable *X*, then the data is binned with bin widths of $x + \Delta x$ (where Δ is some small change in the value of *x*).

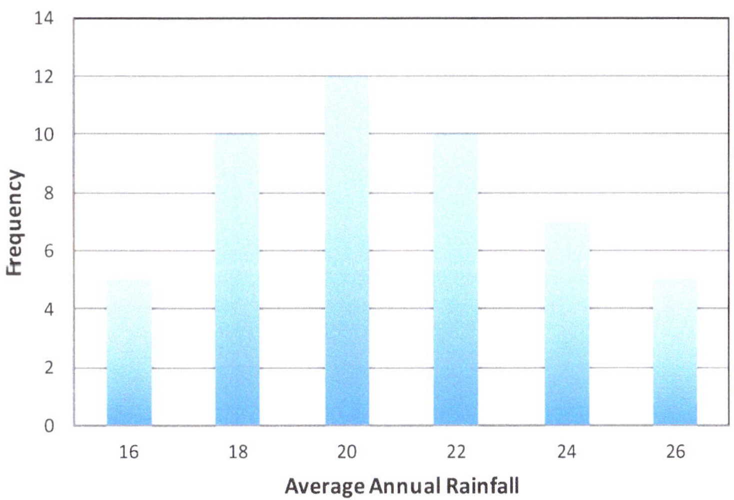

Figure 2.1 Histogram or frequency plot of rainfall data.

‡ Metric units will be used throughout this text, with a preference for SI units. Using SI units is consistent with the rest of the world (only the US still uses English units), helps greatly in minimizing computational errors, and is far easier to use and understand. The goal is that someday the US will move out of the past and adopt the SI system once and for all.

As can be seen in the histogram, the most frequent average annual rainfall value is 20, and 20 happened 12 of the 49 years resulting in 12/49=0.25 or 25% of the time. There are no hard rules about selecting the bin width, however the bin width can have a dramatic impact on what the histogram looks like and what information it conveys. A convention that can be used to determine the bin width is (Benjamin and Cornell, 1970):

$$k = 1 + 3.3 \cdot \log(n) \qquad \textit{equation } 2.1$$

Where k is the number of intervals between the maximum and minimum values observed, n is the number of data observed, and log is the base 10 logarithm.

If we divided the frequency of occurrence, the y-axis on the histogram, by the total number of observations it would then show relative frequency. Relative frequency is useful if we are comparing two different data sets that have different total number of observations; it normalizes the data. We can also sum these relative frequencies at each bin to plot the cumulative relative frequency distribution (Figure 2.2).

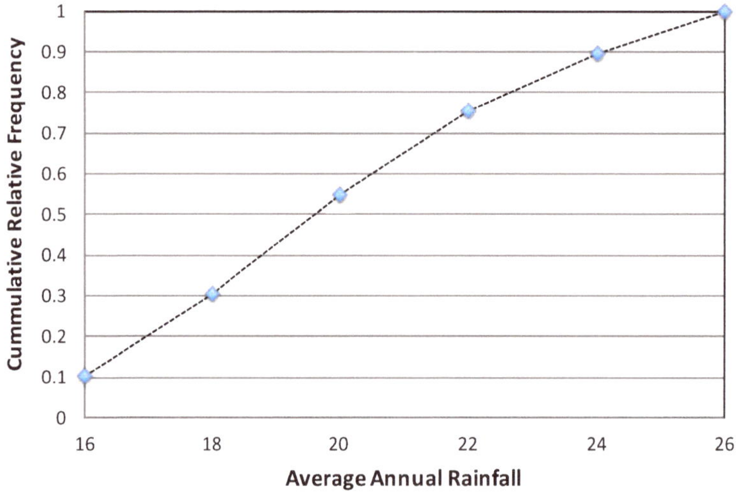

Figure 2.2 Cummulative relative frequency plot of rainfall data.

Both the frequency plot and the cumulative frequency plot (normalized or not) provide us with useful information about the data. The frequency plot shows the most likely outcome and how the data clusters around this value. The cumulative plot shows there is a gradual change between bins and the relative frequency for values $\leq$ x. The general shape of these plots are going to be important in subsequent chapters when we tackle probability and theoretical probability distributions.

Central Tendency

The central tendency is measuring the most likely outcome of a data set. The histogram in Figure 1 shows us the most likely value, visually but that can be influenced by the bin size. The most common measure of central tendency of a data set is the average or arithmetic mean, called the **sample mean**. For a data set of observed values $x = x_1, x_2, \dots, x_n$ the sample mean is:

$$\bar{x} = \frac{1}{n}\sum_{i=1}^{n} x_i \qquad \textit{equation } 2.2$$

If we calculate the sample mean of the rainfall data it is $\bar{x}$=20.77. The **mode** is the most frequently occurring value, which is 20 for the rainfall data as observed in the histogram. The final common measure of central tendency is the **median** which is the value with 50% on either side, or the middle value in an ordered list (the middle value if n is odd, the average of the two middle values if n is even). For the rainfall data the media is $x_{0.5}$=20.

Note that when the frequency plot is symmetric the mean, median, and mode are equivalent (mean=mode=median). This is a nice property of symmetric data or symmetric distributions.

Dispersion

Dispersion is the measure of how the data is dispersed or spread out from the central tendency. This is the variability of a particular problem. The variability is due to aleatory and epistemic uncertainty and has been captured in the measurement of the data set. Ways of describing dispersion include the range, the sample variance, sample standard deviation, and the coefficient of variation.

The **range** is the spread between the maximum and minimum values in a data set. The rainfall data set ranges from 16 to 26, therefore the range is 10. This measure of dispersion is simple but tends to emphasize the extreme values and neglect the bulk of the data around the central tendency. The **sample variance** conceptually is analogous to the moment of inertia about a center of gravity. It is also analogous to the squares of distance. Calculating the sample variance is accomplished by:

$$s^2 = \frac{1}{n-1}\left(\sum_{i=1}^{n} x_i^2 - n\bar{x}^2\right) \qquad \textit{equation } 2.3$$

Dividing by $n-1$ provides an unbiased estimate of the sample variance (Benjamin and Cornell, 1970; Ang and Tang, 2007). The dispersion is usually reported as the **sample standard deviation**, s, which is the square root of the sample variance. The rainfall data has a sample standard deviation of s=2.97. A normalized form of dispersion is the **coefficient of variation**, which is the sample standard deviation divided by the sample mean:

$$\delta = \frac{s}{\bar{x}} \qquad equation\ 2.4$$

This is quite useful when comparing the dispersion from different data sets. It is often reported in literature and can be estimated based on trends or limited data. The coefficient of variation for the rainfall data is $\delta = 0.14$ or 14%.

The equations of the central tendency and dispersion are presented here but in most cases these will be calculated using built in functions in common computation software (e.g., Excel, Matlab, Maple, etc.).

Estimation of Variance

The dispersion or variance can often be estimated or approximated when there is insufficient data to calculate it by using the methods above. Three types of estimation methods include; coefficient of variation published in the literature, the "six sigma" approach, and expert consensus.

For most civil engineering problems there is precedent, meaning that someone has tackled a problem similar enough that it can be used as an analog. We can find **published coefficient of variation** values in the literature in various places depending on the problem. These are not catalogued in any one place but can be found in codes, specifications, or technical journal publications. Table 2.2 below provides a starting point for most geotechnical and structural problems.

Because geotechnical engineering uses natural and not man-made materials, there is depth in the geotechnical literature discussing dispersion and parameter uncertainty. An example of this is the work by Kulhawy and Mayne (1990) where they provide a comprehensive assessment of the coefficient of variation for both lab and *in situ* geotechnical tests. These values were accumulated from many tests over many years and can be applied to any specific problem with some confidence that the dispersion will be reasonably approximated. The coefficient of variation is multiplied by the sample mean of any specific problem to arrive at a defensible sample standard deviation.

Table 2.2: Typical Coefficients of Variation for Geotechnical and Structural Properties (after Harr, 1987)

Type	Parameter	δ (%)
Soil	Porosity[1]	10
	Specific Gravity[2]	2
	Water Content-Clay[3]	13
	Degree of Saturation[3]	10
	Unit Weight[4]	3
	Permeability[5]	90-240
	Compression Index-Sandy Clay[6]	26
	Compression Index-Clay[3]	30
	Friction Angle-Gravel[7]	7
	Friction Angle-Sand[7]	12
Structural Loads	Dead Load[8]	10
	Live Load[8]	25
	Snow Load[8]	26
	Wind Load[8]	27
	Earthquake Load[8]	>100
Structural Resistance	Steel-Tension Member-Yielding[8]	11
	Steel-Tension Member-Tensile Strength[8]	11
	Steel-Compression Beam-Uniform Moment[8]	13
	Steel-Plate/Girder-Flexure[8]	12
	Reinforced Concrete-Grade 60 Steel-Flexure[8]	11
	Reinforced Concrete-Grade 40 Steel-Flexure[8]	14
	Reinforced Concrete-Cast In Place Beam-Flexure[8]	8.0-9.5
	Reinforced Concrete-Short Columns[8]	12-16
	Wood-Compressive Strength[9]	19
	Wood-Flexural Strength[9]	19

[1]Schultze (1972); [2]Padilla & Vanmarke (1974); [3]Fredlund & Dahlman (1972); [4]Hammit (1966); [5]Nielsen et al., (1973); [6]Lumb (1966); [7]Schultze (1975); [8]Ellingwood et al., (1980); [9]Borri et al., (1983)

The "**six sigma**" approach can be used for approximating the dispersion where it can be assumed the data has a relatively symmetric distribution (e.g., previous rainfall histogram). The steps are;

- estimate the mean or median value for a variable,
- conceive of the upper and lower extreme values, and
- divide the range by 6 to get the standard deviation.

This gives an estimate of 99.9% of the data which is ±3 standard deviations on either side of the mean. This is a useful approach, but note that humans are notoriously bad at estimating extreme values. Many failures of Civil projects are due to poor understanding of extreme loads or unforeseen load combinations. So caution must be taken when applying this approach.

The method of **expert consensus** is used in many situations where the dispersion is unquantified and there are many sources that contribute to the uncertainty. This method is just as it sounds, get a group of experts together and have them estimate the dispersion. The experts can be polled to give their best estimate of the standard deviation, or their best estimate of the range which would then lend to the "6 sigma" approach. The same caveat of extreme values applies here as well, experts are also susceptible to a limited ability to conceive of extreme values.

Correlation of Paired Data

Engineering data often comes in pairs. At the beginning of this chapter we discussed the equation of a line where *Y* was a function of the *X*. Here *X* is the independent variable and *Y* the dependent variable where the slope m and the intercept b are treated as coefficients. We infer the relationship between *X* and *Y* through the mathematical function of a straight line. If *X* and *Y* are positively correlated that means that as *X* increases so does *Y*. Negative correlation would indicate the opposite trend, as *X* increases *Y* decreases. We can calculate how *X* and *Y* vary together, this is called the sample covariance:

$$s_{xy} = \frac{1}{n}\sum_{i=1}^{n}(x_i - \bar{x})(y_i - \bar{y}) \qquad equation\ 2.5$$

But more useful is to normalize the sample covariance to arrive at the sample correlation coefficient:

$$\rho = \frac{s_{xy}}{s_x s_y} = \frac{1}{n}\sum_{i=1}^{n}\left(\frac{x_i - \bar{x}}{s_x}\right)\left(\frac{y_i - \bar{y}}{s_y}\right) \qquad equation\ 2.6$$

The correlation coefficient describes the normalized dependence between the two variables, and takes a value in the range from +1 to -1. If there is one to one positive dependence between *X* and *Y* then the correlation coefficient is equal to 1. Figures 2.3 and 2.4 give examples of covariance and correlation coefficient.

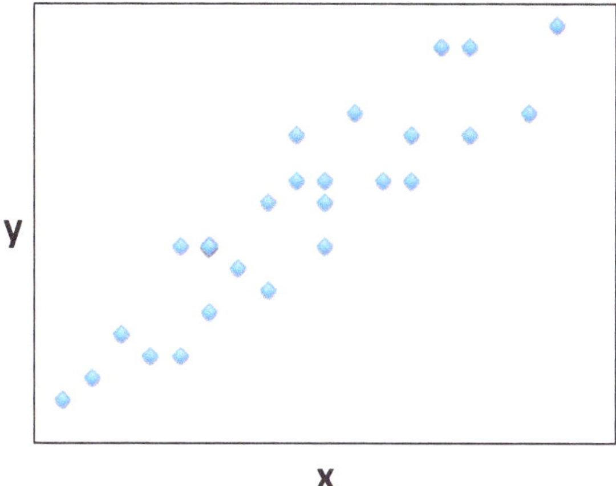

Figure 2.3 Paired data showing small covariance (s_{xy}=small) and positive correlation (ρ~ 1).

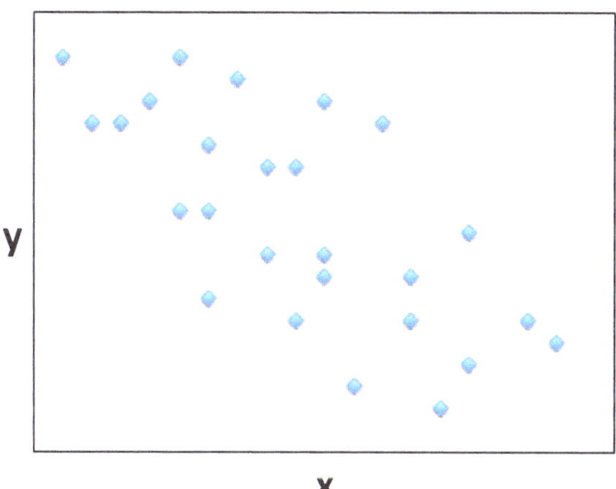

Figure 2.4 Paired data showing large covariance (s_{xy}=large) and negative correlation (ρ~ -1).

A quick means of estimating the correlation coefficient of paired data is by using the properties of an ellipse. As shown in Figure 2.5 the height and the width of the ellipse can be used in Equation 2.7 to estimate ρ. The correlation coefficient is proportional to the ratio of the ellipse width, *h*, to the ellipse height, *H*:

$$\rho = \sqrt{\left(1 - \left(\frac{h}{H}\right)^2\right)} \qquad equation\ 2.7$$

Note as the ratio goes to zero the correlation coefficient goes to 1. This ellipse approximation provides a good check on the correlation found using regression presented the next section.

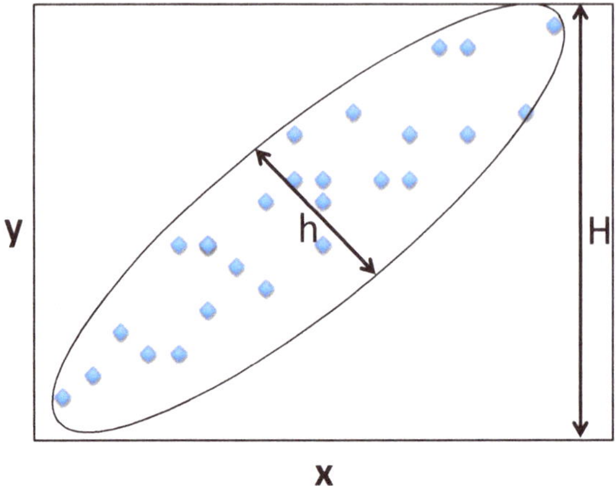

Figure 2.2 Estimating correlation coefficient using an ellipse.

Basic Linear Regression

The most common approach to determine the correlation coefficient in engineering is by linear regression. Spreadsheet programs have "best fit" functions built into the graphing operations making this a simple task. Here only the very basics of regression will be covered so that the correlation value can be used later when we get to probability and systems. There are any number of text books written about linear, multilinear, nonlinear, orthogonal, Bayesian, and other types of regression (e.g., Ang and Tang, 2007; Moss, 2009).

Least squares linear regression is a means of best fitting a straight line to paired data, *X* and *Y*. The objective is to get the mean or expected value of *Y* given $X=x$ using a straight line as the mathematical relationship between the two:

$$E(Y|X = x) = \beta x + \alpha \qquad \textit{equation } 2.8$$

The slope, β, and the intercept, α, are the coefficients we need to solve for given a particular set of observed paired data. A "best fit" straight line is one that minimizes the error for all observations. The total absolute error for all the data points can be represented by the total cumulative squared error:

$$\Delta^2 = \sum_{i=1}^{n} (y_i - \alpha - \beta x_i)^2 \qquad equation\ 2.9$$

The solution for this, a partial differential equation minimizing the squared error in the slope and intercept (Ang and Tang, 2007), is:

$$\beta = \frac{\sum(x_i - \bar{x})(y_i - \bar{y})}{\sum(x_i - \bar{x})^2} \qquad equation\ 2.10$$

$$\alpha = \bar{y} - \beta\bar{x} \qquad equation\ 2.11$$

The variance of this "best fit" is the sample conditional variance of *Y* given *X*:

$$s_{Y|x}^2 = \frac{\Delta^2}{n-2} \qquad equation\ 2.12$$

If we normalize the conditional variance of *Y* given *X* by the variance of *Y* alone we arrive at a measure of the reduction in variance due to *X* because they are correlated and their variances are interrelated:

$$r^2 = 1 - \frac{s_{Y|x}^2}{s_Y^2} \qquad equation\ 2.13$$

The r^2 is a common metric used in regression to qualify how well the line fits the data. A high r^2 indicates that there is a greater reduction in the conditional variance associated with the regression which results in a better prediction of Y. It has been shown (Ang and Tang, 2007) that the correlation coefficient is related to r^2 as the number of samples, n, becomes large:

$$\rho = \sqrt{1 - \frac{n-2}{n-1}\frac{s_{Y|x}^2}{s_Y^2}} \approx \sqrt{r^2} \qquad equation\ 2.14$$

This then brings us back to paired data and the correlation of *X* and *Y*, but here using linear regression to estimate the correlation coefficient. Notice the similarities between Equation 2.14 and Equation 2.7; both are the square root of one minus the ratio of the variance of *Y* given *X* over the variance of *Y* alone.

Example: Shear Strength with Depth

This example (after Moss, 2009, appendix) presents the results of a linear regression analysis. We are interested in how the shear strength changes linearly with depth given the data in the table.

Depth	Shear Strength
6	0.28
8	0.58
14	0.5
14	0.83
18	0.71
20	1.01
20	1.29
24	1.5
28	1.29
30	1.58

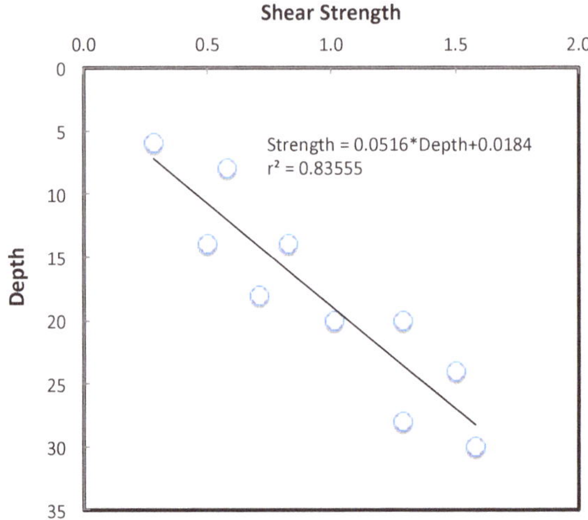

The equations for linear regression are presented in this chapter, but it can be carried out easily using the curve fit option in Excel, or using the curve fitting toolbox in Matlab. The results show a fairly linear relationship between the two variables with an r^2 of 0.84, suggesting that for this problem depth is a reasonably good predictor of shear strength. Note that the variables have been switched in this problem because viewing depth along the vertical axis makes intuitive sense.

Chapter Summary

- A **histogram** is a useful way of viewing the frequency of data and how it is distributed.
- Frequency data can also be plotted in cumulative form and normalized to provide the relative **cumulative frequency**.
- The **central tendency** of data can be described using the **mean**, **median**, and/or **mode**.
- The **dispersion** of data can be described using the **variance**, **standard deviation**, and/or **coefficient of variation**.

- Estimations of variance can be achieved by looking for reported coefficient of variation published in the literature, using the "6 sigma" method, and soliciting expert consensus.
- Paired data can be plotted to determine if correlation exists. A rough estimate of the **correlation coefficient** can be made using the ellipse approach.
- Linear least squares regression can also be used to estimate the correlation coefficient of paired data.

3 ELEMENTARY PROBABILITY AND SET THEORY

We have interrogated the past in Chapter 2 by applying statistical tools to some data. Now in order to design for future events we need to forecast what is likely to happen during the design life of an engineered feature. This is where probability comes in. Probability is heavily used in Civil Engineering risk analysis because as mentioned before data is often scarce.

Set theory is a logical framework for defining the relationships between events. Venn diagrams provide a visual way of defining these relationships and in defining probability. Probability is itself an independent branch of mathematics that is:

- Logically consistent,
- Founded on 3 axioms, and
- Defines the components of probability [but does not define what probability is or what it means].

Probability theory arose from gambling. A friend of Blaise Pascal's (the renown mathematician of the mid-1600's) posed a dice question to him about the probability of winning a particular dice game with certain combinations. Pascal started a technical discussion with his equally renown colleague Pierre de Fermat, and through a series of letters the two worked out the theory of probability (Bernstein, 1998; Gonic and Smith, 1993). The 3 axioms that Pascal and Fermat worked out are still the basis for probability theory today:

Axiom1: For every event (A) there is a probability; $P(A) \geq 0$,
Axiom2: The probability of all events $(S)$is one; $P(S) = 1$,
Axiom3: For two mutually exclusive events, $(A)$and (B), the probability of the union of the two events is the sum of the probabilities of the individual events; $P(A \cup B) = P(A) + P(B)$.

The first two axioms are relatively self explanatory and define the range of probability from 0 to 1. The third axiom with the terms mutually exclusive and union will be defined after a discussion of the nature of probability and set theory.

Nature of Probability

The nature of probability is often vehemently argued in certain circles with the same vigor that is usually reserved for politics or religion. In Civil Engineering we utilize all forms or interpretations of probability to aid in solving our problems. Here probability will be classified into four types:

1. Probability from **relative frequency** (i.e., observed from statistical information). As with the rainfall data in Chapter 2 we observed the most frequent value of 20 occurred 25% of the time (12/49).
2. Probability from ***a priori*** information, usually defined by elementary or geometric constraints of the problem (e.g., when flipping a coin P(tails)=50%).
3. Probability **assigned subjectively**, per expert consensus or engineering judgment.
4. Probability from **mixed information**:
   (observed)+(*a priori*)+(subjective). Using all available information like this is often called a Bayesian approach.

Example: Fault Rupture (*a priori* probability)

A fault has been discovered under the foundation of a power plant. The fault has been mapped at 150km long and the power plant is located 50km from one end. Seismological investigations indicate that the fault is likely to produce a M_W6.5 event which could result in 30km of surface fault rupture contained within the 150km total length. The fault is equally likely to rupture along its entire length. Should the M_W6.5 event occur, what is the probability that the surface fault rupture occurs under the power plant?

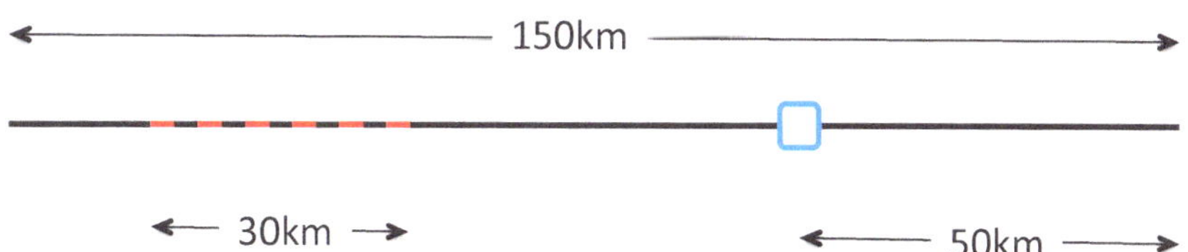

The 30km surface rupture can equally occur anywhere along the fault length, but is contained within the 150km fault (150-30km). We are concerned with the likelihood of it occurring beneath the power plant (30km).

$$P(rupture\ beneath\ plant) = \frac{30}{(150-30)} = \mathbf{0.25}\ or\ \mathbf{25\%}$$

Venn Diagrams and Set Theory

Venn diagrams are an intuitive way of visualizing events and visualizing probabilities. Set theory describes the logical/mathematical relationship between events. The following is the set theory terminology and operators we will be using for events, along with a Venn diagram visually showing the relationships being described.

The Venn diagram below and left shows the entire sample space (S) as represented by the box. The circle (A) represents an area within the sample space that is our event of interest. The sample space (S) is **collectively exhaustive** meaning it contains all possible events or combinations of events, therefore it has a probability of 1. The probability of (A) is equal to the area of (A) with respect to the total area of the sample space (S).

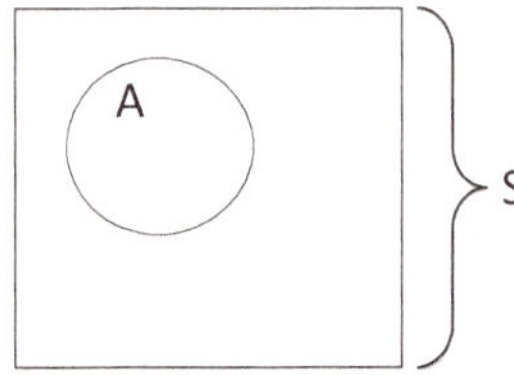

$$P(A) = \frac{P(A)}{P(S)} = \frac{fractional\ area\ of\ A}{1} \qquad equation\ 3.1$$

This concept of area ratios is useful in conceptualizing an event with respect to all possible events for a particular problem. When we have two events that are **mutually exclusive**, meaning that they share no area of sample space in common, then the union is accomplished by adding their fractional areas.

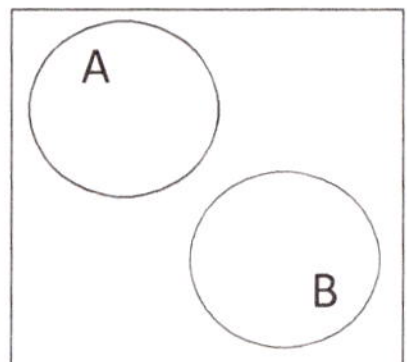

$$P(A \cup B) = P(A) + P(B) \qquad equation\ 3.2$$

Union $(\cup)$

$A \cup B$ is the occurrence of A **or** B or both events. The operative word here is **or** which is the word you want to equate with the union $(\cup)$ symbol. As the Venn diagram shows the union includes the sample space of both events combined. Here the two events are not mutually exclusive because they have some sample space in common.

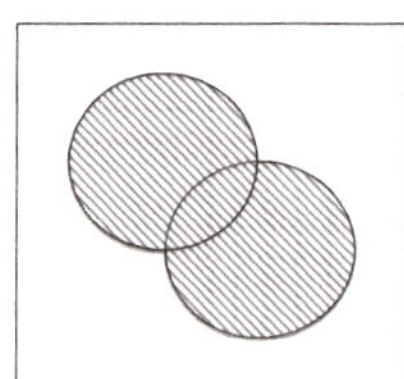

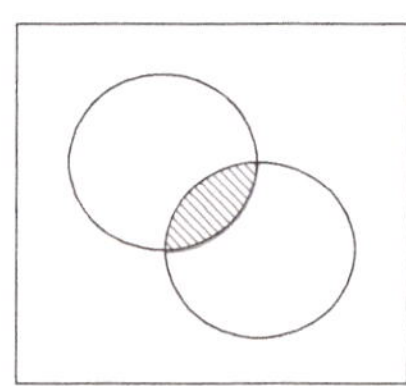

Intersection $(\cap)$

$A \cap B$ is the occurrence of A **and** B jointly. The operative word here is **and** which is the word you want to equate to the intersection $(\cap)$ symbol. As the Venn diagram shows the intersection includes only the area where the two events overlap or intersect. Often the intersection symbol is dropped and the set theory statement becomes simply AB.

Set theory follows the same logical rules as arithmetic, such as the commutative, associative, and distributive rules. The commutative rule means that the order of events in a union or intersection does not affect the outcome, that is:

$$A \cup B = B \cup A \text{ and } AB = BA \qquad equation\ 3.3$$

The associative rule means that all events are equally associated for a union or intersection:

$$(A \cup B) \cup C = A \cup (B \cup C) \text{ and } (AB)C = A(BC) \qquad equation\ 3.4$$

And the distributive rule means that an event can be distributed equally when there is an intersection:

$$(A \cup B)C = (AC) \cup (BC) \qquad equation\ 3.5$$

These rules should be familiar and almost intuitive because there are the same rules we learn at an early age with respect to addition and multiplication.

Compliment

The compliment of an event is one minus the event, it is the sample space that represents everything but the event. The compliment is represented by a bar over the event (e.g., $\bar{A}$). Therefore:

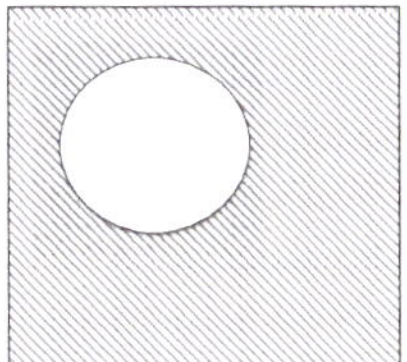

$$P(A) = 1 - P(\bar{A}) \qquad equation\ 3.6$$

$$P(\bar{A}) = 1 - P(A) \qquad equation\ 3.7$$

Addition Rule

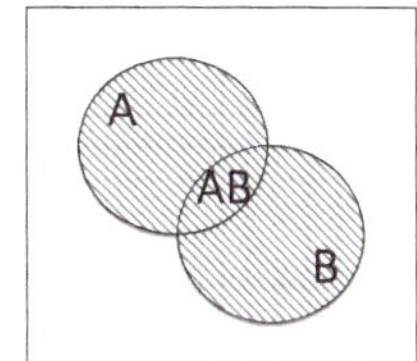

The addition rule defines the mathematics of the union of events. The union of event A and event B includes the combined area of A and B. The union states that the outcome could be event A, **or** event B, **or** both. We sum the area of A and the area of B, but have double counted the intersection so must subtract AB to arrive at the combined area:

$$P(A \cup B) = P(A) + P(B) - P(AB) \qquad equation\ 3.8$$

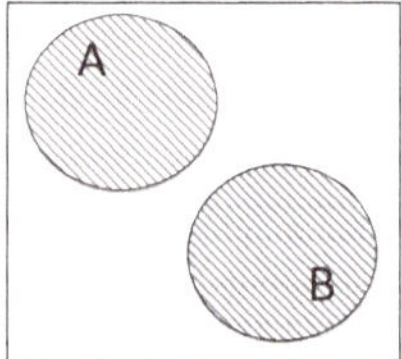

If A and B are **mutually exclusive (ME)** that means there is no joint area and the intersection is zero, therefore:

$$P(A \cup B) = P(A) + P(B) \;\; if\ ME \qquad equation\ 3.9$$

Extending the addition rule for three events results in:

$$P(A \cup B \cup C) = P(A) + P(B) + P(C) \\ -P(AB) - P(BC) - P(AC) + P(ABC) \quad equation\ 3.10$$

Here we subtract the double counted area, but then must add the triple intersection back in to arrive at the union. A Venn diagram can aid in visualizing this solution.

de Morgan's Rule

What is commonly called de Morgan's rule is not a rule itself but more a subset of the addition rule. Nonetheless it is traditionally called de Morgan's rule and will be referred to as such in this text. De Morgan's is an alternate way of solving the addition rule when the compliment is easier to compute. A Venn diagram is often the best way to visualize de Morgan's:

$$P(A \cup B) = 1 - P(\overline{A \cup B}) = 1 - P(\bar{A}\bar{B}) \quad equation\ 3.11$$

$$P(A \cup B \cup \ldots \cup Z) = 1 - P(\overline{A \cup B \cup \ldots \cup Z})$$

$$= 1 - P(\bar{A}\bar{B} \ldots \bar{Z}) \qquad equation\ 3.12$$

The Venn diagram of the union of A and B is shown at the top of the next page:

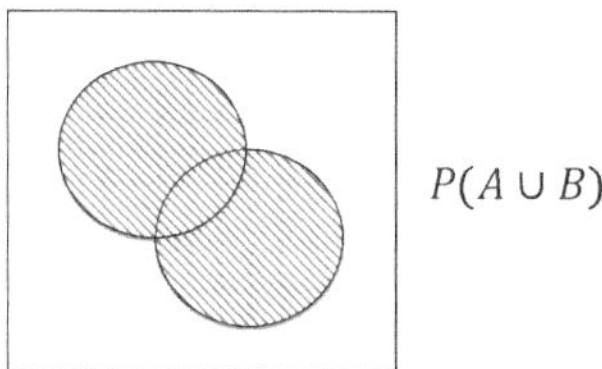

The Venn diagram of the compliment of the union is everything outside the union (see below), and if we take the complement of that we arrive at the first diagram.

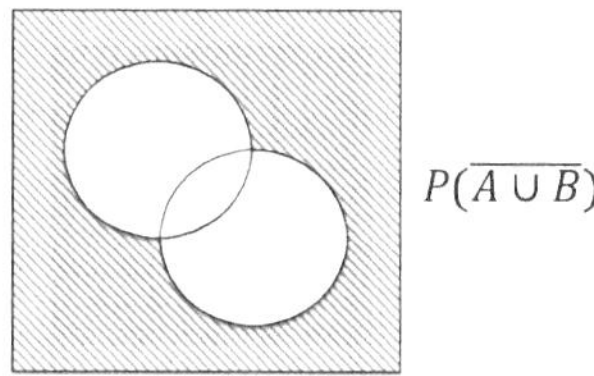

The Venn diagram of the intersection of the complement of A and the complement of B is the double hatched region (see below), and if we take the complement of that we arrive at the first diagram again.

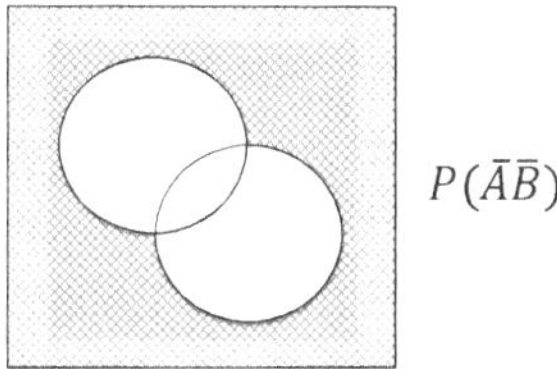

Example: Water and Power Demand

A 6 building complex has recently been constructed and the buildings are unleased as of yet. The engineer wants to meet the demand for water and power utilities as closely as possible without costly over- or under-targeting. For any one building the electricity can be either 5 or 10 units, and the water can be either 1 or 2 units. The total sample space then looks like:

E_5W_2	$E_{10}W_2$
E_5W_1	$E_{10}W_1$

The owner ascribes the following probabilities (expert consensus) based on previous experience with similar type of buildings:

$$\left.\begin{aligned} P(E_5W_1) &= 0.1 \\ P(E_{10}W_1) &= 0.1 \\ P(E_5W_2) &= 0.2 \\ P(E_{10}W_2) &= 0.6 \end{aligned}\right\} \text{sums to 1.0} \checkmark$$

(a) What is the probability of water demand of 2 units?

$$P(W_2) = P(E_5W_2) + P(E_{10}W_2) = 0.2 + 0.6 = \mathbf{0.8}$$

Note that $P(E_5W_2)$ and $P(E_{10}W_2)$ are mutually exclusive.

(b) What is the probability of power demand of 10 units?

$$P(E_{10}) = P(E_{10}W_1) + P(E_{10}W_2) = 0.1 + 0.6 = \mathbf{0.7}$$

(c) What is the probability of either the water demand is 2 units or the power demand is 10 units?

$$\begin{aligned} P(W_2 \cup E_{10}) &= P(W_2) + P(E_{10}) - P(W_2 \cap E_{10}) && \textit{addition rule} \\ &= P(W_2) + P(E_{10}) - P(E_{10}W_2) && \textit{intersection} \\ &= \;\; 0.8 \;\; + \;\; 0.7 \;\; - \;\; 0.6 \;\; = \mathbf{0.9} \end{aligned}$$

Note: In solving these problems I will annotate to the right what rule was used for each step to aid in problem solving clarity.

Conditional Probability Rule

It's been said that every probability is a conditional probability, and this statement rings true the more one spends working on probability problems. Conditional probability defines the measure of dependence between events; it is large when events occur jointly, small when they don't, and zero when events are mutually exclusive. A typical conditional probability question is "What is the probability of event A given the occurrence of event B?":

$$P(A|B) = \frac{P(AB)}{P(B)} \qquad \textit{equation } 3.13$$

Because B has occurred the sample space is limited to the area of A within B; the area is restricted to the conditional event space. The area of A occurring within B is the intersection of A and B. The conditional probability is this intersection normalized by B, the ratio of the intersection AB to the total area of B.

If B has no influence on A then we can say that the two events are **statistically independent** (SI).

$$P(A|B) = P(A)\ if\ SI \qquad equation\ 3.14$$

The above statement (Equation 3.14) is the only means of determining if two events are statistically independent. Statistical independence pertains to the influence of one event on another, which is much different from mutual exclusivity that pertains to the sample space two events have in common. Events A and B may intersect but that does not dictate whether they are statistically independent or not.

Another way to think of conditional probability is through the correlation coefficient introduced in Chapter 2. In both we are describing the influence of one event upon another, however the correlation coefficient is constrained to a linear relationship.

$$\begin{aligned} if\ P(A|B) > P(A)\ then\ \rho = + \\ if\ P(A|B) < P(A)\ then\ \rho = - \\ if\ P(A|B) = P(A)\ then\ \rho = 0 \end{aligned}$$

When the conditioned event has a higher probability than the marginal event then there is a positive correlation, when it has a lower probability than the marginal event then there is a negative correlation, and when the conditioned event has an equal probability to the marginal event the events are statistically independent and the correlation coefficient is zero.

Note that often $P(A)$ and $P(A|B)$ come from an engineering study and the joint probability $P(A \cap B)$ or $P(AB)$ is what is desired, where:

$$\begin{aligned} P(AB) &= P(A|B)P(B) \\ &= P(B|A)P(A) \end{aligned}$$

Example: Water and Power Demand II

What's the probability that a building that needs 10 units of power will also require 2 units of water? In this problem the probability of 2 units of water is conditioned on the demand for 10 units of power.

$$P(W_2|E_{10}) = \frac{P(E_{10}W_2)}{P(E_{10})} = \frac{0.6}{0.7} = \mathbf{0.86} \quad \textit{conditional probability rule}$$

Example: Rolling Dice

Two die are rolled together. What is the probability the dice sum to 3?

1 1
2 2
3 3 2/36
4 4
5 5
6 6

Now one die is rolled and comes up 1. What is the probability that both dice sum to 3?

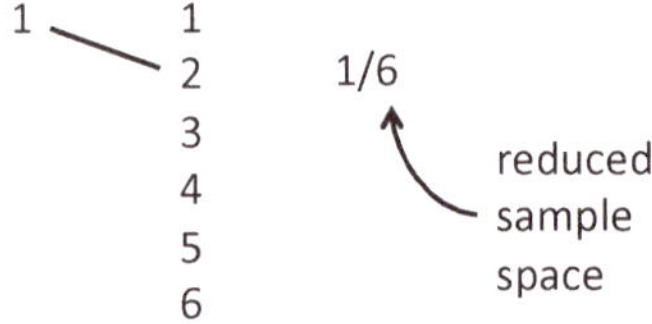

In the second situation the sample space is reduced because it is conditional.

Multiplication Rule

The multiplication rule defines the mathematics for the intersection of events. We rearrange the conditional probability rule to solve for the intersection or joint probability:

$$P(AB) = P(A|B)P(B) \qquad \textit{equation } 3.15$$

If we determine that A and B are **statistically independent** then the conditional probability statement is reduced and we have the multiplication of two marginal event probabilities:

$$P(AB) = P(A)P(B) \ for \ SI \qquad equation \ 3.16$$

For two events we can reverse the statement to check statistical independence:

$$P(A)P(B) = P(AB) \qquad equation \ 3.17$$

But this does not necessarily hold true for more than two events (Der Kiureghian, 2001) and one should always check statistical independence with the conditional probability rule. For three events the multiplication rule becomes:

$$P(ABC) = P(A|BC)P(B|C)P(C) \qquad equation \ 3.18$$

Example: Building Foundation

The foundation of a tall building may fail due to inadequate bearing capacity (B) or excessive settlement (S). The following probabilities are known for this type of building and failure situation:

$P(B) = 0.001;\ \ P(S) = 0.008;\ \ P(B|S) = 0.10$

(a) What is the probability of failure of the foundation? Failure would be B or S or both.

$$\begin{aligned} P(B \cup S) &= P(B) + P(S) - P(BS) && addition\ rule \\ &= P(B) + P(S) - P(B|S)P(S) && conditional\ probabilty\ rule \\ &= 0.001 + 0.008 - 0.10(0.008) = \mathbf{0.0082} \end{aligned}$$

(b) What is the probability the building will experience excessive settlement but not bearing capacity failure?

$$\begin{aligned} P(S \cap \bar{B}) = P(S\bar{B}) &= P(S|\bar{B})P(\bar{B}) && multiplication\ rule \\ &= P(\bar{B}|S)P(S) && commutative\ property \\ &= [1 - P(B|S)]P(S) && compliment \\ &= [1 - (0.10)]0.008 = \mathbf{0.0072} \end{aligned}$$

Example: Fault Rupture II

Readdressing the 150km long fault with the potential for 30km of surface fault rupture. For this example there are three pipes that cross the fault.

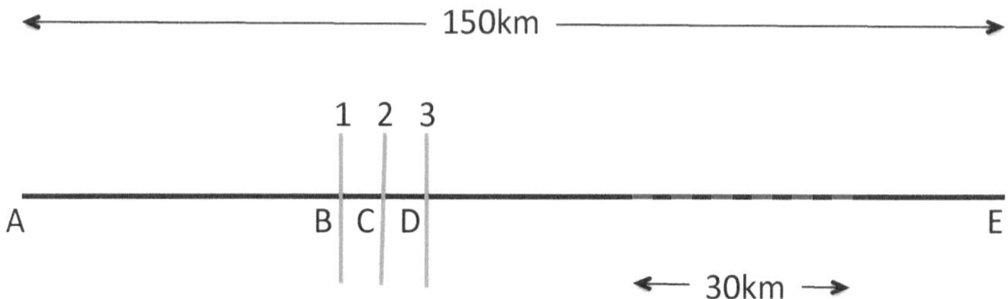

The distance between points: A to B = 60km, B to C = 10km, C to D = 12.5km, and D to E = 67.5km. P_i here denotes a failure of pipe *i*.

(a) What is the probability of failure of each pipe? The solution is the same as for the power plant.

$$P(P_i) = \frac{30}{(150-30)} = \mathbf{0.25} \quad a\ priori$$

(b) What is the probability of P_1 and P_2? The 30km surface rupture must include the 10km distance between B and C.

$$P(P_1 \cap P_2) = P(P_1P_2) = \frac{(30-10)}{(150-30)} = \mathbf{0.167} \quad a\ priori$$

(c) What is the probability that 1 breaks, given that 2 has broken?

$$P(P_1|P_2) = \frac{P(P_1P_2)}{P(P_2)} \quad conditional\ probabilty\ rule$$

$$= \frac{0.167}{0.25} = \mathbf{0.667}$$

(d) Are P_1 and P_2 statistically independent?

$P(P_1|P_2) \neq P(P_1)\ therefore\ not\ SI$

(e) What is the probability of 1 and 3 breaking?

$$P(P_1P_3) = \frac{(30-10-12.5)}{(150-30)} = \mathbf{0.0625} \quad a\ priori$$

(f) What is the probability of at least one of the first two pipes failing?

$$\begin{aligned} P(P_1 \cup P_2) &= P(P_1) + P(P_2) - P(P_1P_2) \quad addition\ rule \\ &= 0.25 + 0.25 - 0.167 = \mathbf{0.333} \end{aligned}$$

(g) The first two pipes are redundant and only one needs to survive to maintain continued service (water, phone, other...). What is the probability at least one will survive?

$$\begin{aligned} P(\bar{P}_1 \cup \bar{P}_2) &= 1 - P(P_1P_2) \quad de\ Morgan's\ rule \\ &= 1 - 0.167 = \mathbf{0.8333} \end{aligned}$$

(h) What is the probability that both will survive?

$$\begin{aligned} P(\bar{P}_1\bar{P}_2) &= 1 - P(P_1 \cup P_2) \quad de\ Morgan's\ rule \\ &= 1 - [P(P_1) + P(P_2) - P(P_1P_2)] \quad addition\ rule \\ &= 1 - [0.25 + 0.25 - 0.167] = 1 - 0.333 = \mathbf{0.667} \end{aligned}$$

Problem Solving Rubric

Solving elementary probability problems can be intuitive for some people and opaque for others. However, as the complexity of the problems increases they tend to become opaque to all. In order to aid in solving elementary probability problems there are some commonalities that lend to a routine problem solving rubric. Practice of this rubric with simple problems makes solving more complex problems tractable and routine.

Rubric

1) **Declare all knowns.** This can be tricky in some problems because the knowns may be implicit, inobvious, or not directly stated.
2) **Write the problem statement in set terminology.** This means translating the word statement into unions and/or intersections of events. This is often the hardest part of any elementary probability problem and takes careful dissection of the problem statement and translation of it into set theory.

3) **Reduce the statement to a calculable form by invoking the rules of probability**. If the set theory statement is correct then this step is often a straight forward manipulation of the rules of probability until a mathematical calculation can be performed.

This rubric works for most problems of elementary probability, but not all. There are a few things to watch out for when problems get tricky:

- In certain problems *a priori* probability is part of the knowns but not explicitly stated.
- Some problems result in complex probability statements that can be simplified by invoking de Morgan's rule.
- When in doubt draw a Venn diagram to clarify what the probability statement is describing.
- Adhere to logic when solving these problems.
- Always perform a "gut check" on the answer when a resulting probability is calculated. Is the answer reasonable and does it answer the word statement?

Total Probability Theorem

Total probability is a theorem that applies to a marginal event probability that is composed of many conditional events. Equation 3.19 and the Venn diagram below are shown for three condition events but this theorem can be extended to any number of conditional events, Equation 3.20.

$$P(A) = P(A|E_1)P(E_1) + P(A|E_2)P(E_2) + P(A|E_3)P(E_3) \quad equation\ 3.19$$

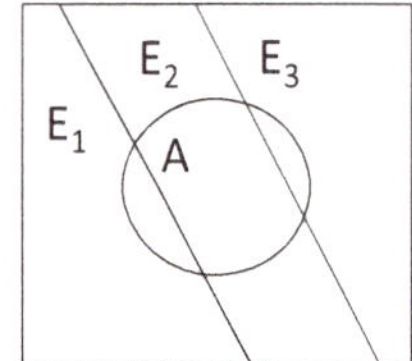

$$P(A) = \sum_{i=1}^{n} P(A|E_i)P(E_i) \quad equation\ 3.20$$

The best way to elucidate this theorem is through example.

Example: Hurricane Damage

Hurricanes are categorized by increasing wind speed, Category 1 through Category 5. Based on historical hurricane data along the Louisiana Coast the annualized probability of each category is;

$P(C_1) = 0.35, P(C_2) = 0.25, P(C_3) = 0.14, P(C_4) = 0.05, P(C_5) = 0.01.$

In this problem we are interested in the probability of structural damage due to hurricane winds. Reconnaissance of previous hurricane disasters have shown that structural damage can be approximated by the following conditional probabilities;

$P(D|C_1 \dots C_5) = 0.05, 0.10, 0.25, 0.60, 1.00$

Given this information, what is the annual probability of structural damage?

$$
\begin{aligned}
P(D) &= P(D|C_1)P(C_1) \dots P(D|C_5)P(C_5) \quad \textit{total probability theorem} \\
&= 0.05 \cdot 0.35 + 0.10 \cdot 0.25 + 0.25 \cdot 0.14 + 0.60 \cdot 0.05 + 1.00 \cdot 0.01 \\
&= \mathbf{0.1175}
\end{aligned}
$$

Bayes' Theorem

If we combine the conditional probability rule and the multiplication rule we arrive at what is called Bayes' Theorem.

$$P(B|A) = \frac{P(A|B)P(B)}{P(A)} \qquad \textit{equation } 3.21$$

This theorem (often called Bayes' Rule) is a simple enough statement that has a lot of history behind it and has caused more than a few scandals in the realm of statistics and probability. It is named after Thomas Bayes, a minister who worked on probability in his spare time. He perished before publishing his work but his proofs were published posthumously by his colleague Richard Price circa 1763. The theorem was independently rediscovered and published by Simon-Pierre Laplace (1812).

Bayes' theorem describes an inverse probability, that of event B given the occurrence of event A. In this form the theorem is often used for updating the probability of something given subsequent information. The following examples demonstrate this Bayesian Updating.

Example: Aggregate Supply

In this problem there is aggregate being delivered to a construction site from two sources, A and B. Trucks from sources A deliver 600 loads a day of which 3% is bad (meaning it does not meet the project specifications). Trucks from source B deliver 400 loads a day of which 1% is bad.

(a) What is the probability of bad aggregate?

First we must declare the probability that the aggregate is from each source and declare the conditional probability of bad aggregate.

$$P(A) = \frac{600}{(600 + 400)} = 0.60 \quad P(B) = \frac{400}{(600 + 400)} = 0.40$$

$$P(Bad|A) = 0.03 \quad P(Bad|B) = 0.01$$

$$P(Bad) = P(Bad|A)P(A) + P(Bad|B)P(B) \quad total\,probability$$
$$= 0.03 \cdot 0.60 + 0.01 \cdot 0.40 = \mathbf{0.022}$$

(b) If the aggregate is bad, what is the probability it was from source A?

This is the type of inverse probability statement that lends well to using Bayes' theorem. Often in problems like this total probability and Bayes' theorem go hand-in-hand.

$$P(A|Bad) = \frac{P(Bad|A)P(A)}{P(Bad)} \quad Bayes'theorem$$

$$= \frac{0.03 \cdot 0.60}{0.022} = \mathbf{0.818}$$

Example: Pollution Control

This problem deals with urban air quality. It has been determined that a city's air pollution is primarily due to two sources; industrial (*I*) and auto (*A*) emissions. In the next 5 years the chances of controlling these two emissions sources have been estimated at 75% and 60% respectively. If only one of the two sources are controlled, there is an 80% probability of bringing air pollution under control (*C*).

(a) What is the probability of controlling air pollution?

The probability of controlling air pollution is conditioned on controlling the two sources. It can be divided into four conditional probability statements that are summed together per total probability.

If both sources are controlled then control is obviously achieved, 100%. As stated previously if one of the other sources is controlled, the probability of air quality control is 80%. And if neither source is controlled then there is no chance of control, 0%. In this problem we are assuming that industrial and auto emissions are statistically independent.

$$P(C) = P(C_1|AI)P(AI) + P(C_2|A\bar{I})P(A\bar{I}) + P(C_2|\bar{A}I)P(\bar{A}I) + P(C_3|\bar{A}\bar{I})$$
$$= 1(0.60 \cdot 0.75) + 0.80(0.60 \cdot 0.25) + 0.80(0.40 \cdot 0.75) + 0(0.4 \cdot 0.25)$$
$$= \mathbf{0.81}$$

(b) If pollution is not controlled after 5 years, what is the probability it was due entirely to auto pollution?

All the conditional and marginal probabilities needed to answer this inverse probability statement can be found in the total probability statement above.

$$P(\bar{A}I|\bar{C}) = \frac{P(\bar{C}|\bar{A}I)P(\bar{A}I)}{P(\bar{C})} \quad Bayes' theorem$$

$$= \frac{0.20 \cdot 0.30}{0.19} = \mathbf{0.32}$$

Books, theses, and other documents have been written on the philosophical underpinnings of Bayes' Theorem and applications using a Bayesian approach (e.g., McGrayne, 2011). The Bayesian approach has been used quite successfully in areas where there is limited data and an estimate of the probability must be made. This procedure of estimating the probability given limited data however tends to fly in the face of classical or frequentist statistics and for a period of time, spanning the early to middle 1900's, was considered equivalent to statistical heresy. Nonetheless people found it quite useful for solving difficult problems where there was no classical solution, and it gradually took hold. Since the 1960's on there was movement to provide a robust mathematical basis for the Bayesian approach which succeeded in establishing the validity on equal footing with classical statistics. To this day there are still intellectual battles between the two camps, classical vs. Bayesian, but this often has more to do with ones own philosophical view of determinism than which method solves the problem adequately.

Chapter Summary

- Probability is founded on three axioms.
- A probability can be of different forms depending on the problem and the way it is framed; relative frequency, *a priori*, subjective, and Bayesian.
- Set theory is a way of logically expressing the **union** (**or**) or **intersection** (**and**) of events.
- Venn diagrams are a means of visually showing how events are related within the sample space.
- Tools for solving probability problems are the rules and theorems that are an extension of the three axioms of probability; **compliment**, **addition rule**, **de Morgan's rule**, **conditional probability rule**, **multiplication rule**, **total probability theorem**, and **Bayes' theorem**.
- Using a problem solving rubric is useful as probability problems become more complex and non-intuitive. The rubric provides a systematic way of solving most problems encountered in probability.
- Two events are **mutually exclusive** when they share no sample space in common. When mutually exclusive, the occurrence of one event precludes the occurrence of the other.
- **Statistical independence** is when the occurrence of one event does not influence the probability of occurrence of another. The dependence between events is measured using the conditional probability rule.

4 RANDOM VARIABLES AND PROBABILITY DISTRIBUTIONS

A random variable is a mathematical tool for describing an event. It maps the possible outcomes of an event onto a number line as shown below. The Venn diagram shows the sample space with events A and B. These are mapped onto a number line to aid in mathematical calculations. In Figure 4.1, A and B are mutually exclusive, whereas in Figure 4.2 they have some sample space in common.

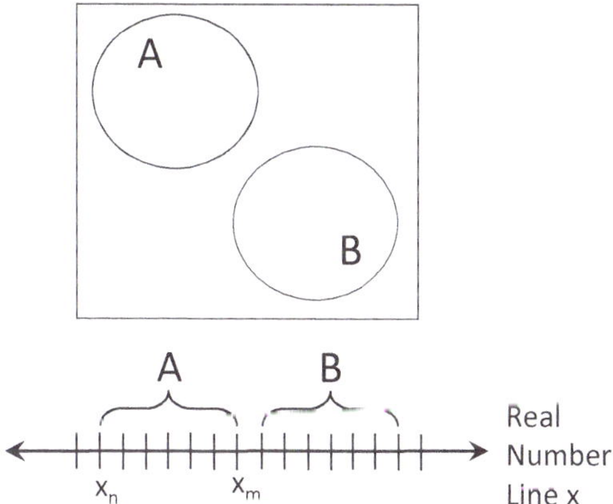

Figure 4.1 Mapping mutually exclusive events to a real number line.

The random variable in these figures is denoted by a capital letter, X, and the number that the random variable assumes is the lower case, x. Here event A is mapped to the random variable X and is comprised of all real numbers within that space: $A = \{x_n \leq X \leq x_m\}$.

We can look at the range of values that X can assume and calculate the relative probability of any particular value or range of values, this then defines a probability distribution.

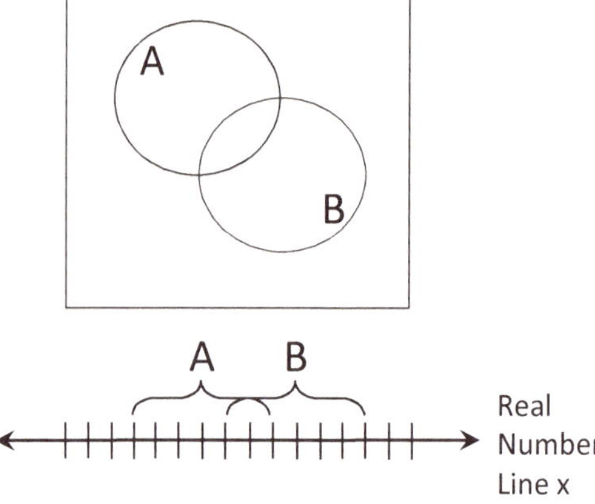

Figure 4.2 Mapping not-mutually exclusive events to a real number line.

Probability Distribution

A distribution is classified first by the type of number it describes, whether it is discrete or continuous. An example of a discrete number in Civil Engineering would be a traffic count (e.g., 12 cars made a left hand turn at the intersection). An example of a continuous number would be a contaminant concentration (e.g., 0.055 grams/cubic meter of polyvinyl chloride).

The mathematical density of a discrete probability distribution is shown in Figure 4.3. This is analogous to a histogram, showing the likelihood of any discrete value occurring with respect to all other discrete values. A discrete probability distribution is called a probability mass function, **PMF**, is denoted by the symbol $p_X(x)$ and is equal to the probability of the random variable X assuming a particular value $P(X = x)$ for all x values.

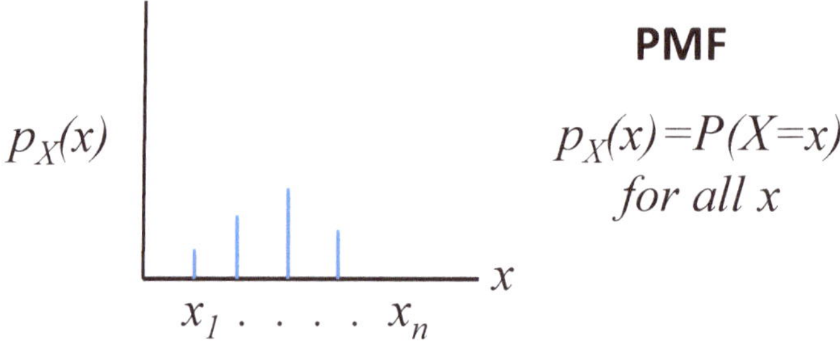

Figure 4.3 The probability mass function, PMF, is the probability distribution for a discrete number.

The cumulative distribution of a discrete number is called a **CDF** or cumulative distribution function. It is denoted by the symbol $F_X(x)$ and is equal to the probability of the random variable X being less than or equal to a particular value:

$$F_X(x) = \sum_{all\; x_i \leq x} P(X = x_i) = \sum_{all\; x_i \leq x} p_X(x_i) = P(X \leq x) \quad equation\ 4.1$$

Figure 4.4 shows an example of a cumulative distribution function. Notice that the cumulative probability sums to 1.0 at the upper bound.

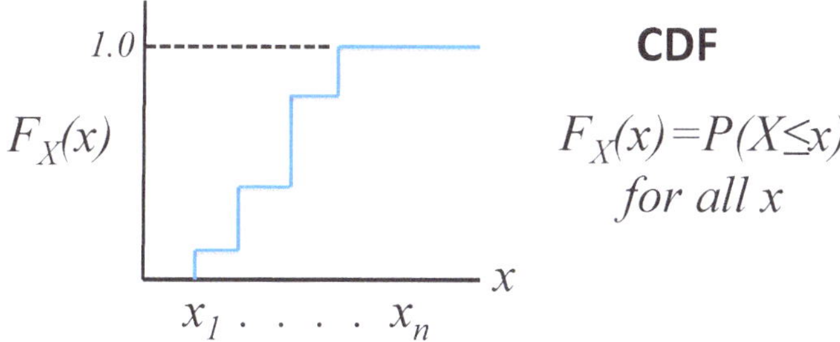

Figure 4.4 The cumulative distribution function, CDF, for a discrete number.

For continuous numbers we must define a small interval dx to describe the probability density because for a continuous number $P(X = x) = 0$ as the number goes out to an infinite number of decimal places. The probability distribution of a continuous number is called the probability density function, **PDF**, is denoted by the symbol $f_X(x)$ and is equal to the probability of the random variable over the small interval $P(x < X \leq x + dx)$. Figure 4.5 shows a probability density function, of which the area under the entire curve must sum up to 1.0.

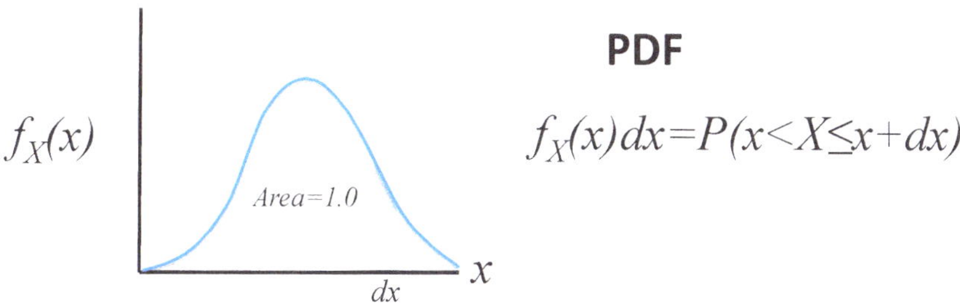

Figure 4.5 The probability density function, PDF, for a continuous variable.

We can also show the cumulative distribution function, **CDF**, for a continuous variable in the same manner as before, but instead of summing we integrate.

$$F_X(x) = \int P(x \leq X \leq x + dx)$$

$$= \int_{-\infty}^{x} f_X(x)dx = P(x \leq X) \qquad equation\ 4.2$$

Figure 4.6 shows the cumulative distribution function for a continuous variable, with the upper bound approaching 1.0.

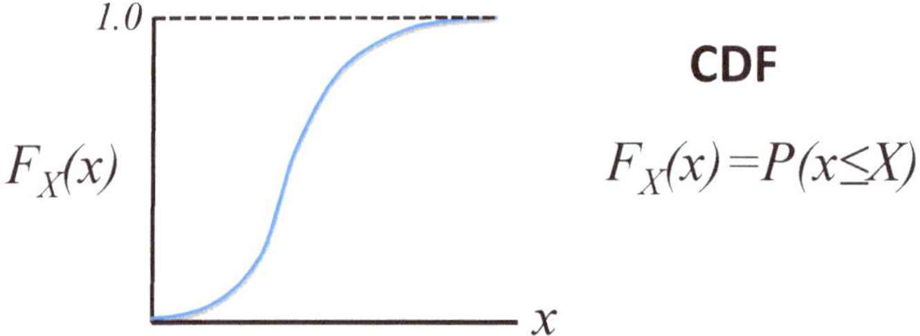

Figure 4.6 The cumulative distribution function, CDF, for a continuous variable.

By defining probability distributions in this manner we can then answer probability questions with ease. If we are dealing with a discrete number, like train crossings, and want to know how many crossings have occurred then we consult the PMF for that specific number. If we have a continuous number, like axial load on a column, and want to know the probability of exceeding some load then we consult the CDF for that number and take the compliment. We will discuss more problem solving of this type later in the chapter.

Generalities of Probability Distributions

Probability distributions can take on any mathematic form (i.e., any equation) as long as they satisfy the three axioms of probability; the function must be nonnegative and probabilities of all possible values must add up to 1.0. To frame this with a CDF:

1) $F_X(-\infty) = 0$ and $F_X(\infty) = 1.0$,
2) $F_X(x) \geq 0$ for all x, and is nondecreasing with x,
3) $F_X(x)$ is continuous to the right.

Therefore any mathematical function satisfying the three axioms is a valid probability distribution. There are fundamentally two reasons for the existence of a probability distribution; mathematical utility (e.g., Normal distribution), and conceptual utility (e.g., Poisson distribution). An entire menagerie of predefined probability distributions exist that people have come up with to solve some problem requiring the mathematical or conceptual utility of those particular functions. These can be found in statistics text or reference books (e.g., Ang and Tang, 2007). But to reiterate, any mathematical function that satisfies the three axioms is a valid probability distribution. Examples of common probability distributions:

- Continuous - Uniform (i.e., a straight line), Triangular, Exponential, Normal, Lognormal, Gamma, Beta, Extreme value…
- Discrete – Binomial, Poisson, Geometric…

In this text we will only be using the Normal and Lognormal, for mathematical ease and simplicity, but all the discussions herein apply to any probability distribution.

Some things to keep in mind when selecting and applying a probability distribution to a particular problem:

- Does the variable of interest represent something that is infinite or does it have a finite upper bound?
- Is it a variable that can assume a negative value, or must it be positive?
- To fit a distribution to a particular problem some tricks that are often used; shifted distributions, compound or mixed distributions, or truncated and renormalized continuous distributions.

There is nothing magical about any particular mathematical function that is a probability distribution, you simply want one that best describes the phenomenon of interest and doesn't introduce any more epistemic uncertainty than necessary. That being said the choice of a distribution is often subjective and should be thoughtfully and clearly justified.

Note: It is common to drop the subscripts on the generic names of probability distributions for brevity. Therefore:

$$PMF \quad p_X(x) = p(x)$$
$$PDF \quad f_X(x) = f(x)$$
$$CDF \quad F_X(x) = F(x)$$

Example: Wind Loading

A structural engineer is designing a tall tower to withstand wind loads. Based on the maximum annual wind velocity recorded over many years the histogram of the data looks like the figure below. The engineer decided to model the data using a negative exponential function because it provides a reasonable fit to the data, a subjective decision that introduced epistemic uncertainty to the problem but provided mathematical convenience in solving the problem.

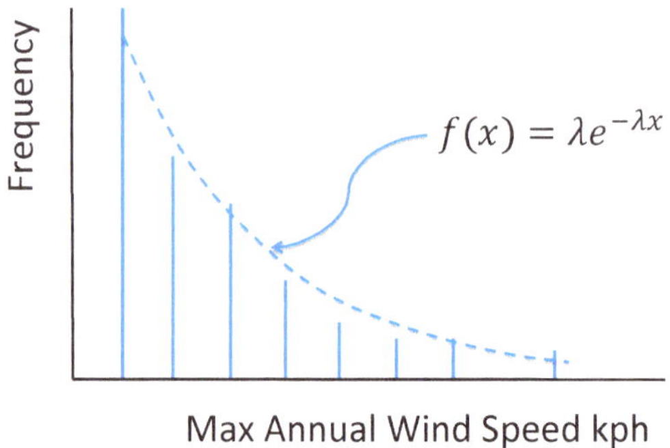

The negative exponential function is now being used as a PDF describing the probability of max annual wind speed $P(x < X \leq x + dx)$.

(a) What is the CDF for this problem?

The CDF is the integral of the PDF and will allow us to answer questions about less than or greater than a particular value easily.

$$F(x) = \int_0^x f(x)dx = \int_0^x \lambda e^{-\lambda x} dx = -e^{-\lambda x} \mid_0^x = 1 - e^{-\lambda x} \; where \; x \geq 0$$

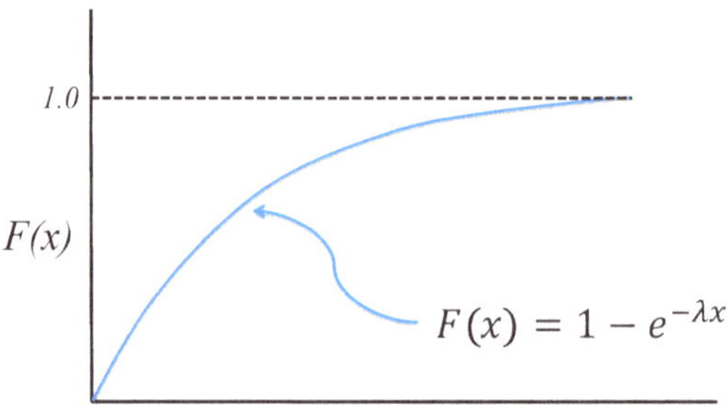

(b) Data shows that the probability of the max annual wind speed less than 70 kph is 90%. Estimate the coefficient (λ) of the function.

We could use the PDF and integrate from 0 to 70, or the CDF at that value since it is the integral of the PDF.

$0.9 = P(0 \leq X \leq 70) = P(X \leq 70) = F(x)$

$0.9 = 1 - e^{-\lambda 70}$ $\qquad e^{-\lambda 70} = 0.1$

$-\lambda 70 = \ln(0.1)$ $\qquad \lambda = \mathbf{0.033}$

Now that we have the PDF, the CDF, and the coefficient in the equation we can move forward to answer any number of engineering related probability questions.

(c) What's the probability of the max annual wind speed between 35 and 70 kph?

Here we want the area under the PDF from 35 to 70 kph. With the CDF this is accomplished by evaluating at 70 and then subtracting at 35 because the CDF is the pre-integrated area under the PDF.

$P(35 \leq X \leq 70) = F(70) - F(35)$

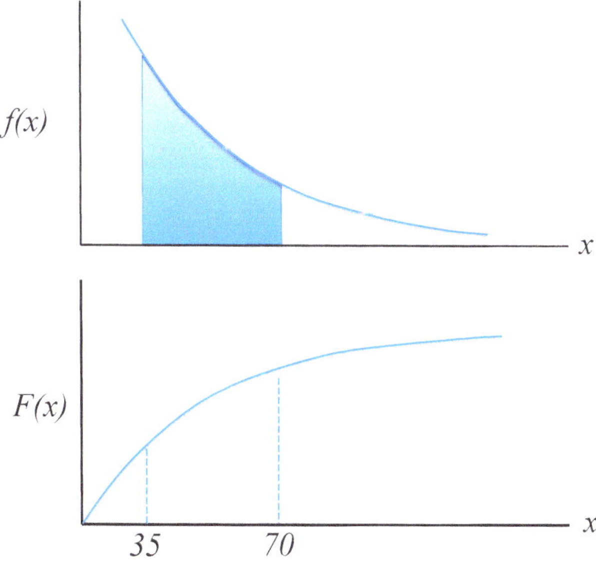

$$P(35 \leq X \leq 70) = \left(1 - e^{-0.033(70)}\right) - (1 - e^{-0.033(35)})$$

$$= 0.900 - 0.685 = \mathbf{0.215} \text{ or } \mathbf{21.5\%}$$

(d) What is the probability that the max annual wind speed will be greater than 140kph?

The area under the PDF that is greater than 140kph answers this question, which is equivalent to the compliment of the value of the CDF at 140kph.

$$P(X \geq 140) = 1 - P(X \leq 140) = 1 - F(140)$$

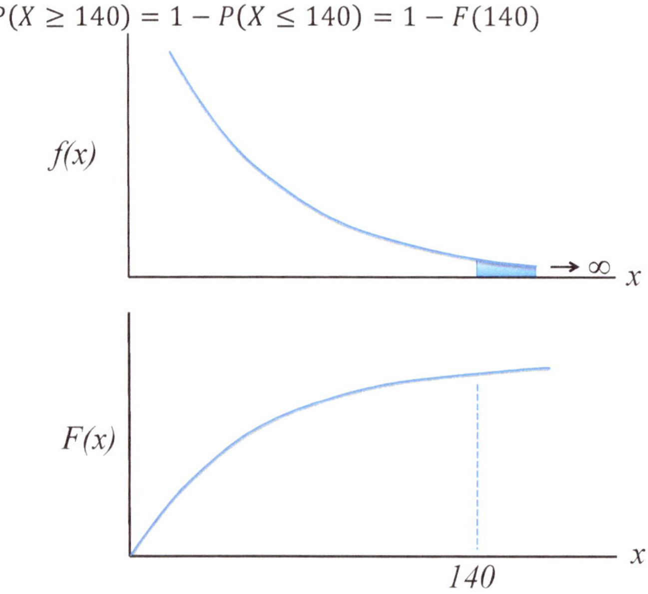

$P(X \geq 140) = 1 - \left(1 - e^{-0.033(14)}\right) = \mathbf{0.0099}$ or approximately **1%**

Expectation and Moments

A probability distribution is a theoretical representation of the frequency of occurrence, analogous to the histograms generated from data as shown in Chapter 2. We can quantify the central tendency and dispersion of probability distributions just the same as we did with sample statistics from empirical data.

Central Tendency

The central tendency can be quantified as the mean, mode, or median. The **mean** for probability distributions, called the **expectation** or **first moment**, for discrete and continuous numbers respectively are:

$$E(x) = \mu = \sum_{i=1}^{n} x_i \cdot p(x_i) \qquad equation\ 4.3$$

$$E(x) = \mu = \int_{-\infty}^{\infty} x \cdot f(x)dx \qquad equation\ 4.4$$

The **mode** is the most probable value or value of highest probability and is denoted by $\tilde{x}$. The **median**, as before with sample statistics, is the value with 50% on either side, or when the CDF is equal to 50%:

$$x_m = F(x_m) = 50\% \qquad equation\ 4.5$$

When a probability distribution is symmetric the mean, median, and mode are equivalent.

Dispersion

The spread of a probability distribution can be quantified by the **variance** for discrete and continuous respectively as:

$$Var(x) = \sum_{i=1}^{n} (x_i - \mu)^2 \cdot p(x_i) \qquad equation\ 4.6$$

$$Var(x) = \int_{-\infty}^{\infty} (x - \mu)^2 \cdot f(x)dx \qquad equation\ 4.7$$

The **standard deviation**, often called the **second moment**, is the square root of the variance.

$$\sigma = \sqrt{Var(x)} \qquad equation\ 4.8$$

The normalized form of dispersion is the **coefficient of variation**, which is the standard deviation divided by the mean:

$$\delta = \frac{\sigma}{\mu} \qquad equation\ 4.9$$

Higher moments can be calculated but are generally unimportant for the purposes of engineering risk analysis. The equations of the moments are presented here but in most cases these will be calculated using built in functions in common computation software (e.g., Excel, Matlab, Maple, ect.).

Multivariate Probability Distribution

We may encounter an engineering problem where we are mapping an event onto two lines to describe two random variables involved in the event. We are not limited to two random variables, but with two we can draw the joint probability in three dimensions. For two variables we can define the joint CDF, as well as the joint PMF for discrete numbers and joint PDF for continuous numbers:

$$CDF: F\ (xy) = P(X \leq x, Y \leq y) \quad equation\ 4.10$$

$$PMF:\ p(xy) = P(X = x, Y = y) \quad equation\ 4.11$$

$$PDF: f(xy)dxdy = P(x < X \leq x + dx, y < Y \leq y + dy)\ equation\ 4.12$$

The joint PDF can look something like Figure 4.7, a three dimensional surface showing the joint probability of X and Y.

Multivariate distributions, of coarse, follow the rules of probability. The multiplication rule is the joint probability distribution of X and Y which is equal to the conditional probability distribution of X and Y multiplied by the marginal distribution of Y:

$$f(xy) = f(x|y)f(y) \quad equation\ 4.13$$

If X and Y are statistically independent random variables then the multiplication rule simplifies to the product of their marginal probability distributions:

$$f(xy) = f(x)f(y) \quad equation\ 4.14$$

The marginal distribution can be determined from a joint distribution by integrating out the other marginal distribution:

$$f(y) = \int_{-\infty}^{\infty} f(xy)dx \quad equation\ 4.15$$

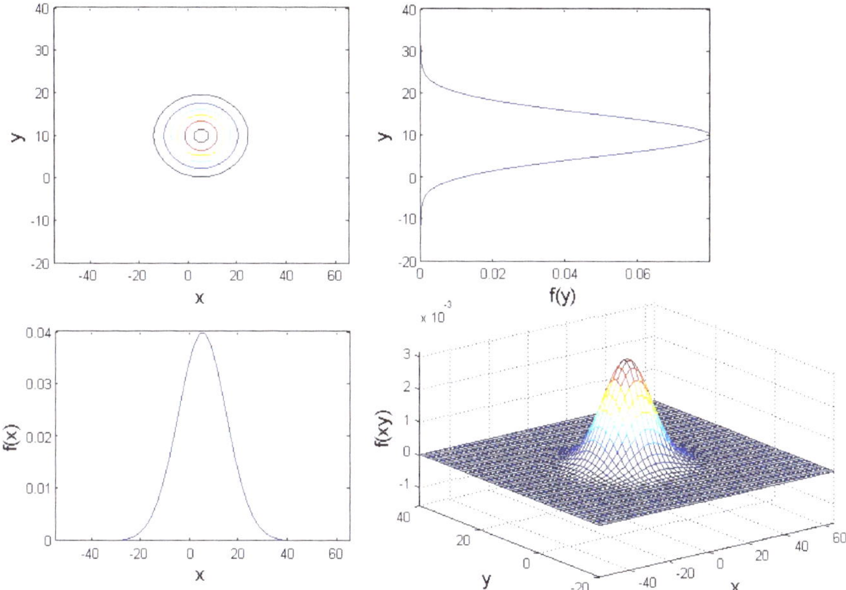

Figure 4.7 Joint PDF of X and Y showing their marginal distributions, the joint plan view (upper left), and the joint 3D view (lower right).

Similarly all the other rules and theorems of probability apply and can be used to help solve engineering problems. If we are dealing with a joint distribution where the marginals are correlated then we often describe the amount of correlation by the correlation coefficient (as discussed in previous chapters) here applied to probability distributions:

$$\rho = \frac{Cov(xy)}{\sigma_x \sigma_y} \qquad equation\ 4.16$$

$$Cov(xy) = E(xy) - E(x)E(y) \qquad equation\ 4.17$$

Theoretical Distributions: Normal and Lognormal

As mentioned there is an entire menagerie of theoretical distributions that achieve some mathematical utility, conceptual utility, or both. In this text we will confine our discussion of specific theoretical distributions to the normal and lognormal distributions because they satisfy both utilities and will provide us with a quick first-order approximation for the majority of problems encountered in Civil Engineering.

Normal Distribution

The normal distribution, or Gaussian distribution as it is sometimes called, is defined by the following mathematical function:

$$f(x) = \frac{1}{\sigma\sqrt{2\pi}} exp\left(-\frac{1}{2}\left(\frac{x-\mu}{\sigma}\right)^2\right) \quad for\ -\infty < x < \infty \qquad equation\ 4.18$$

This function is symmetric, has a "bell shaped" curve, and is completely defined by the first (μ) and second (σ) moments. The normal distribution has been found to describe the likelihood of many natural phenomena from a very broad range of fields. The history of the normal distribution starts with gambling in the 1700's, traverses through many mathematical treatises for 100's of years, and continues today where it finds almost ubiquitous application in realms utilizing probability and mathematics. The shorthand notation used for the PDF of the normal distribution is $N(\mu, \sigma)$.

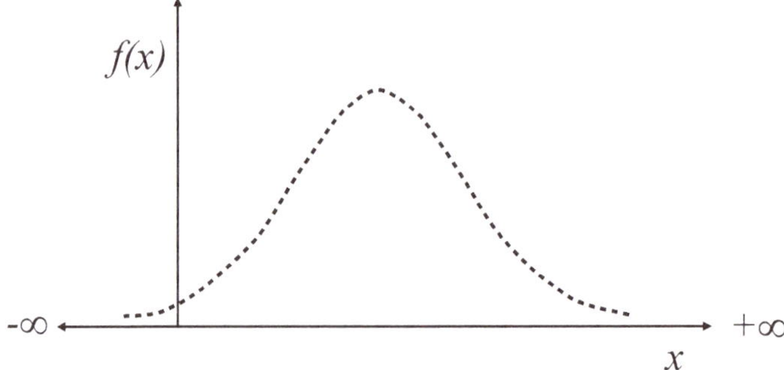

Figure 4.8 Example PDF of the normal distribution, $N(\mu, \sigma)$.

A change in the mean (μ) value will shift the distribution left or right, but leave the breadth of the curve unchanged. A change in the standard deviation (σ) will change the breadth of the curve thereby altering the height of the curve, but leave the central tendency unshifted.

We can also define what is called the standard normal distribution, a normalized distribution that has a mean of zero ($\mu = 0$) and a unit standard deviation ($\sigma = 1$) which can be shorthanded to $N(0,1)$. The indefinite integral of the standard normal distribution, that is the CDF, is denoted by the symbol $\Phi(x)$.

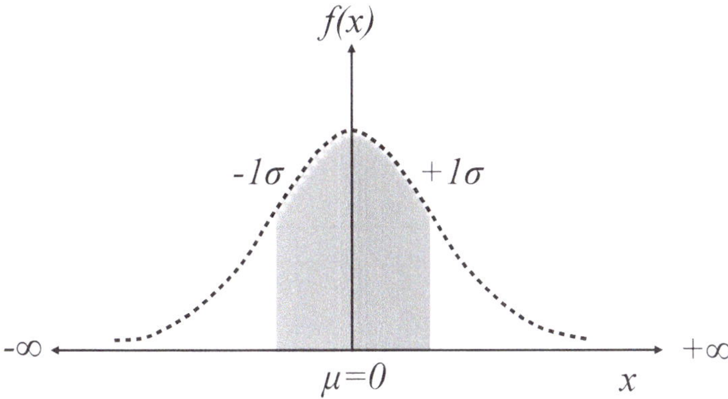

Figure 4.9 The standard normal distribution $N(0,1)$.

Unfortunately there is no closed-form solution to this indefinite integral, therefore the CDF of the standard normal distribution is usually presented in tabular form as can be found in Appendix A. There are also approximation equations of the CDF, also presented in Appendix A. For computational purposes the tabular or approximate CDF of the standard normal distribution is programmed into almost every handheld calculator and computational software available today (e.g., Matlab, Maple, Excel). The tabulated values are given only for positive values because the standard normal CDF is symmetric, therefore:

$$\Phi(-x) = 1 - \Phi(x) \qquad equation\ 4.19$$

And the values of x for probabilities less than $p<0.5$ are negative and can be calculated as:

$$x = \Phi^{-1}(p) = -\Phi^{-1}(1-p) \quad equation\ 4.20$$

For the standard normal distribution the probability (or area) within ± 1 standard deviation is 68.3%, within ± 2 standard deviations is 95.4%, and ± 3 standard deviations is 99.7%. One can see that 3 standard deviations on either side of the mean accounts for almost all of the likelihood for a problem dealing with a normally distributed random variable.

If we have a random variable that is normal (meaning that it's probability distribution is described by the normal distribution), or we can reasonably assume that it is normal, we can easily solve probability problems with the tabulated standard normal distribution results:

$$P(a < X \leq b) = \Phi\left(\frac{b-\mu}{\sigma}\right) - \Phi\left(\frac{a-\mu}{\sigma}\right) \quad equation\ 4.21$$

It has been shown that the sum of independent random variables asymptotically approaches the normal distribution regardless of the distribution of the underlying variables. This property of the normal distribution will become important in the next chapter where we deal with functions of random variables. The above statement affirms the Central Limit Theorem (CLT) which states that the sum of N random variables approaches normality as N becomes large:

$$z = k_1 + k_2 + \cdots + k_n$$

$$z = \sum k \rightarrow N(\mu_z, \sigma_z)$$

Example: River Flows

A river stage-gage (in former Yugoslavia) measured peak annual flows over 39 years. The data appears to be normally distributed with N(28,676; 21,117). Compute the probability that the peak flow will be less than 100,000; 80,000; and 50,000 in any given year, and determine the return period for each of these flows.

$P(X \leq x) for\ N(\mu, \sigma)\ is\ \Phi(x) therefore:$

$$P(X \leq 100{,}000) = \Phi\left(\frac{100{,}000 - 28{,}676}{21{,}117}\right) = \Phi(3.38) \cong 0.99964$$

Where we are using Appendix A to approximate $\Phi(x)$ values.

$$P(X \leq 80{,}000) = \Phi\left(\frac{80{,}000 - 28{,}676}{21{,}117}\right) = \Phi(2.43) \cong 0.99245$$

$$P(X \leq 50{,}000) = \Phi\left(\frac{50{,}000 - 28{,}676}{21{,}117}\right) = \Phi(1.01) \cong 0.84375$$

The return period (T) is usually given as the annual probability of exceedance in years, and exceedance is the compliment of a value less-than-or-equal-to.

$$T = \frac{1}{P(X > x)} = \frac{1}{1 - P(X > x)} = \frac{1}{1 - \Phi(x)} \quad therefore:$$

$$T(100{,}000) \cong \frac{1}{1 - 0.99964} = 2777.8\ yrs$$

$$T(80{,}000) \cong \frac{1}{1 - 0.99245} = 132.5\ yrs$$

$$T(50{,}000) \cong \frac{1}{1 - 0.84375} = 6.4\ yrs$$

So for the lowest flow value (50,000) there is a probability of 84% that the flows will be less than this value, and the return period for this flow is roughly six and a half years.

Lognormal Distribution

The lognormal distribution is just that, the log of the normal distribution. Here log refers to the natural log (ln) or log base e. The mathematical function that describes the lognormal distribution is:

$$f(x) = \frac{1}{\xi x\sqrt{2\pi}} exp\left(-\frac{1}{2}\left(\frac{\ln(x) - \lambda}{\xi}\right)^2\right) \quad for\ x \geq 0 \qquad equation\ 4.22$$

The lognormal distribution starts at zero and goes to infinity which makes this distribution often useful for describing numbers in engineering that cannot take a negative value (e.g., penetration resistance, flow rate).

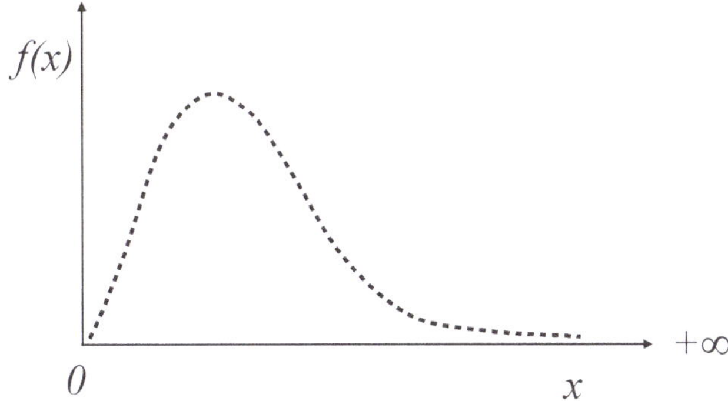

Figure 4.10 Example PDF of the lognormal distribution, $LN(\lambda, \xi)$.

The moments for this distribution are the log transformed first and second moments:

$$\lambda = ln(\mu) - \frac{1}{2}\xi^2 \quad equation\ 4.23$$

$$\xi^2 = ln\left(1 + \frac{\sigma^2}{\mu^2}\right) \quad equation\ 4.24$$

For the lognormal distribution it is often more convenient to use the median and coefficient of variation because:

$$\lambda = ln(x_m) \quad equation\ 4.25$$

$$\xi \approx \delta \ \ for\ \delta \leq 0.3 \quad equation\ 4.26$$

A lognormal distribution with a median of $x_m = 10$ has a $\lambda \cong 2.303$, and the coefficient of variation δ to ξ is one to one for values less than about 30%. Since the lognormal is just the natural log of the normal then we can use the standard normal distribution once we log transform the parameters:

$$P(a < X \leq b) = \Phi\left(\frac{ln(b) - \lambda}{\xi}\right) - \Phi\left(\frac{ln(a) - \lambda}{\xi}\right) \quad equation\ 4.27$$

Because of the summation properties of the normal distribution and the mathematics of logs, it can be shown that the product of N random variables asymptotically approaches the lognormal distribution. This will also be important in the next chapter.

$$z = k_1 k_2 \ldots k_n$$

$$ln(z) = ln(k_1) + ln(k_2) + \cdots + ln(k_n)$$

$$ln(z) = \sum ln(k_i) \rightarrow LN(\lambda_z, \xi_z)$$

Example: Rainfall Intensity

The average annual rainfall over a 24hr period in San Luis Obispo in the Fall months is $\bar{x}$ = 2.3 cm. A rather intense storm came through on 10/13/2009 that dropped 8.8cm within 24hr. What is the probability that there could be a storm with a higher 24hr total? Since the rainfall total cannot be less than zero we will assume that the distribution is lognormal. The shaded area in the figure represents the question we are asking.

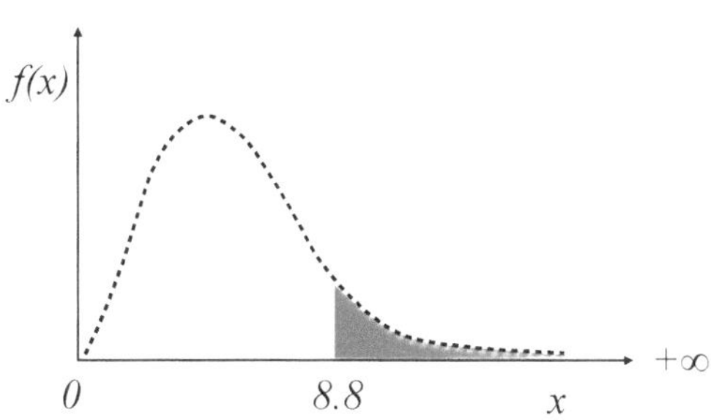

There is no information on the variability, only the average was reported by the weather service. Here we will use the "6 sigma" approach to make an estimate of the uncertainty to arrive at a rough first-order estimate of the probability. For this problem we will assume that the maximum intensity could be something like 15cm and the minimum would be no rain or 0cm. Dividing this range by 6 gives a sample standard deviation of s = 2.5cm.

Solving for the lognormal parameters given the sample mean and estimated sample standard deviation:

$$\xi^2 = ln\left(1 + \frac{\sigma^2}{\mu^2}\right) \approx ln\left(1 + \frac{s^2}{\bar{x}^2}\right) = \ln\left(1 + \frac{2.5^2}{2.3^2}\right) = 0.78$$

$$\xi = \sqrt{0.78} = 0.88$$

$$\lambda = ln\mu - \frac{1}{2}\xi^2 \approx ln\bar{x} - \frac{1}{2}\xi^2 = ln(2.3) - \frac{1}{2}(0.78) = 0.44$$

With the parameters of the lognormal distribution we can now solve for the probability of exceedance.

$$P(X > 8.8) = 1 - P(X \leq 8.8) = 1 - \Phi\left(\frac{lnx - \lambda}{\xi}\right)$$

$$= 1 - \Phi\left(\frac{ln8.8 - 0.44}{0.88}\right) = 1 - \Phi(1.97) \approx 1 - 0.975$$

$$= \mathbf{0.025}$$

So the first-order answer is that there is a 2.5% chance that a storm can dump more rain than the one that occurred on 10/13/2009. This is of course a function of our assumptions. The maximum assumed value may be underestimated, and if this is so we are estimating a lower probability. The assumption that it follows the lognormal distribution which is infinite in the positive direction gives us a higher probability. But given the minimal information we started with, we were able to estimate that this storm was a relatively intense storm and that the odds of a more intense one are not high.

Example: Structural Supports

A structure has three supports A, B, and C. The loads on these supports can be estimated with reasonable accuracy but the soil conditions are heterogeneous below the support footings. Assume the settlements S_A, S_B, and S_C can be defined as independent normal variates based on subsurface field testing with:

$$\mu_{S_A} = 3.0cm \;\; \delta_{S_A} = 0.20$$
$$\mu_{S_B} = 2.5cm \;\; \delta_{S_B} = 0.30$$
$$\mu_{S_C} = 3.0cm \;\; \delta_{S_C} = 0.25$$

What is the probability that the total settlement exceeds 3.5cm, which would result in the slab of the structure being misaligned with the incoming underground utilities?

$$P(S > 3.5) = P(S_A > 3.5 \cap S_B > 3.5 \cap S_C > 3.5)$$

$$= P(S_A > 3.5)P(S_B > 3.5)P(S_C > 3.5) \quad \textit{statistically independent}$$

$$= \big(1 - P(S_A \leq 3.5)\big)\big(1 - P(S_B \leq 3.5)\big)\big(1 - P(S_C \leq 3.5)\big) \; \textit{compliment}$$

$$= \left(1 - \Phi\left(\frac{3.5 - \mu_A}{\sigma_A}\right)\right)\left(1 - \Phi\left(\frac{3.5 - \mu_B}{\sigma_B}\right)\right)\left(1 - \Phi\left(\frac{3.5 - \mu_C}{\sigma_C}\right)\right)$$

$$\textit{standard normal with}\; \sigma = \delta \cdot \mu$$

$$= \big(1 - \Phi(0.83)\big)\big(1 - \Phi(1.33)\big)\big(1 - \Phi(0.67)\big)$$

$\cong (0.20)(0.09)(0.25) = \mathbf{0.005}$

The probability of any of the supports exceeding 3.5cm is roughly 0.5%, essentially negligible from a settlement standpoint.

Chapter Summary

- A random variable is an event mapped to a real number line. This allows for mathematical calculations of likelihood.
- A probability distribution shows all possible values a random variable can assume and the relative likelihoods of those values.
- Numbers can be discrete or continuous and the probability distributions reflect this. For discrete numbers there is the probability mass function, **PMF**, for continuous numbers there is the probability distribution function, **PDF**, and the sum or integral of these result in the cumulative distribution function, **CDF**. Each of these distributions refer to a specific relationship between the number line and the likelihood of a number or range of numbers.
- Any mathematical function can be a probability distribution as long as it satisfies the three axioms of probability. There are a number of common predefined probability distributions, of which we will be focusing on the normal and lognormal distribution.
- The **first** and **second moment** of a distribution refers to the **mean** and **standard deviation** of the distribution.
- When two or more random variables are of interest we consider a multivariate distribution, where commonly the correlation or independence of the marginal distributions is the primary concern.
- The normal and lognormal distributions satisfy both mathematical and conceptual utility for most Civil Engineering problems and can be used to quickly make a first-order assessment of the probability.
- No closed-form solution for the integral of the normal distribution exists so it is common to use tabulated results of the standard normal distribution $N(0,1)$ in its CDF form $\Phi(x)$ found in Appendix A.

5 FUNCTIONS OF RANDOM VARIABLES: ERROR PROPAGATION

This chapter contains the core of engineering risk analysis, the central theme about which the rest of the material pivots. In Civil Engineering almost all designs involve some calculation using an equation that is based on the physics of the problem, empirical data of the phenomenon, or a combination of the two (often called semi-empirical equations). The parameters in these equations are routinely treated as deterministic and its common to use mean or median values for calculation purposes. But if we treat these parameters as random variables and propagate the uncertainty through the equations we get a much better understanding of the most likely answer, as well as an understanding of how much confidence we should have in that most likely answer. Through this process of error propagation we fully characterize the problem by accounting for the uncertainty of the input variables and their mathematical interrelationship as described in the engineering equation. This sets the stage for determining how accurate the answer is, how much confidence we can have in the mean or median, and how best to proceed to ensure a reliable engineering design.

The phrase **functions of random variables** can be defined as "equations, formula, or mathematical models that contain parameters with uncertainty." The application of this can be demonstrated through the following simple example. Let's say we have the following equation:

$$z = x + y$$

If the input parameters are treated as deterministic, meaning they have no uncertainty or more likely they have uncertainty but it is neglected, then we can simple add the values to determine the results. Let's say that $x = 5$ and $y = 4$, then the solution is $z = 9$. But if we include the uncertainty of the parameters, treating them as random variables, and the probability distributions of the random variables are similar to those shown in Figure 5.1, then the solution is not evident. How do we sum the discrete distribution of X and the continuous distribution of Y to get the resultant distribution of Z?

Should the resultant distribution be discrete or continuous? How are the probabilities combined?

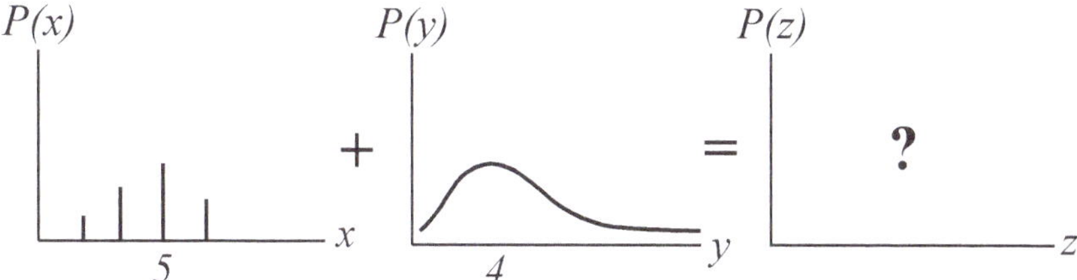

Figure 5.1 A function of random variables, $Z = X + Y$.

It turns out that the solution depends on:

a) The type of mathematical function,
b) The specific distributions of the random variables, and
c) How much information we need of the resultant.

There are three groups of solutions for functions of random variables depending on the answers to a, b, and c above. The three groups are:

1. Exact Solutions,
2. Approximate Solutions, and
3. Computational Solutions.

The box below shows how these three groups map out as presented in the subsequent discussion.

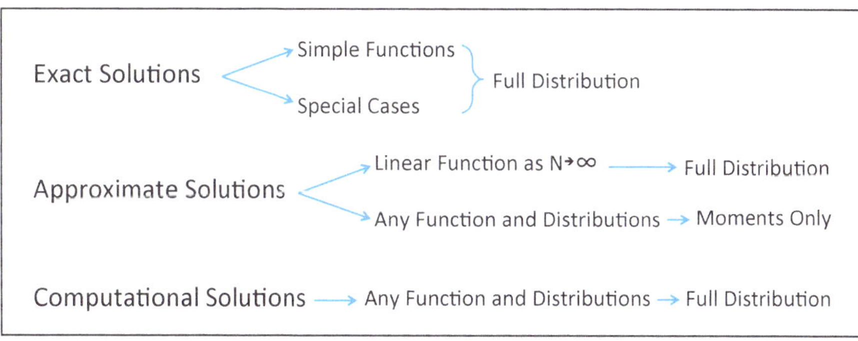

Exact Solutions

Exact solutions will provide the full distribution of the resultant but only work when we have a simple function or a special case. Simple functions are one-to-one functions with single roots. Special cases are when the function is; 1) a sum or difference of normally distributed random variables, or 2) a product or quotient of lognormally distributed random variables.

Simple Function

An example of a one-to-one function with a single root is:

$$Y = X^2 \; where \; x \geq 0$$

In this example there is a direct relationship between the probability of X and the resulting probability of Y. Take the values given in the table below. If we map this we see the direct relationship.

X	P(x)	y	P(y)
1	0.25	1	0.25
2	0.50	4	0.50
3	0.25	9	0.25
4	0	16	0

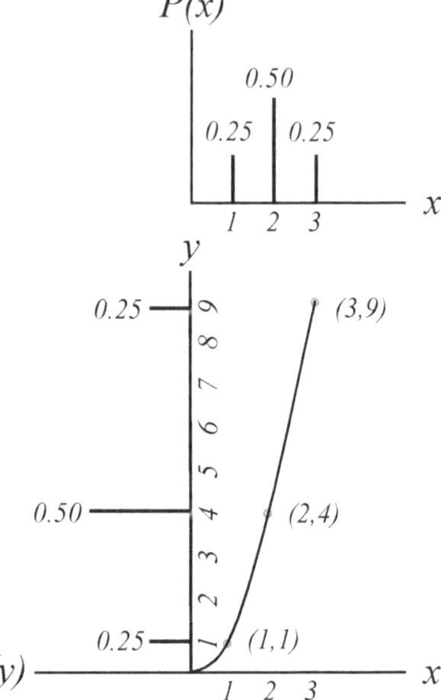

For simple functions such as this there is a one to one relationship of probability, $p(y) = p(x)$. If we had a continuous number then $f(y) = f(x)$. The CDF's would follow accordingly. Unfortunately there are few situations in engineering where we are working with a simple one-to-one function with a single root.

Special Cases

The special cases are a byproduct of the properties of the normal distribution. The first special case is a function that is a sum or difference of normally distributed random variables. If we return to the function we were initially evaluating:

$$Z = X + Y$$

But now X and Y are normally distributed random variables with their respective moments, $N(\mu_x \sigma_x)$ and $N(\mu_y \sigma_y)$, then:

$$\mu_z = \mu_x + \mu_y \qquad equation\ 5.1$$

From this it can be said that "the mean of the sums is the sum of the means." The squared second moment is:

$$\sigma_Z^2 = \sigma_x^2 + \sigma_y^2 + 2\rho\sigma_x\sigma_y \qquad equation\ 5.2$$

These results show that the resultant Z is a normally distributed random variable with moments, $N(\mu_Z, \sigma_Z)$. If the random variables of the input parameters are statistically independent, that is $\rho = 0$, then the squared second moment (a proof be found in detail in Appendix B) reduces to:

$$\sigma_Z^2 = \sigma_x^2 + \sigma_y^2 \qquad equation\ 5.3$$

So when the input parameters are normally distributed and the function is a sum we see the solution is simply the sum of the means and the sum of the variances. We can expand this to any number of random variables along with their coefficients to broaden the application to any sum or difference of Gaussian random variables.

$$Z = \sum_{i=1}^{n} a_i Q_i \qquad equation\ 5.4$$

Where the a's are coefficients (fixed deterministic values) and the Q's are normally distributed random variables. Then generically:

$$\mu_Z = \sum_{i=1}^{n} a_i \mu_{Q_i} \qquad equation\ 5.5$$

$$\sigma_Z^2 = \sum_{i=1}^{n} a_i^2 \sigma_{Q_i}^2 + \sum_{i,j=1}^{n} \sum_{i \neq j}^{n} a_i a_j \rho_{Q_i Q_j} \sigma_{Q_i} \sigma_{Q_j} \qquad equation\ 5.6$$

Note when we have a difference equation the coefficient of -1 is squared in the first term and becomes a +1 having no effect, but does influence the second term that accounts for the correlation. The 2 that we see in front is because of the double sum of the i's and j's (for example when there are 2 terms we would see the combinations of 1,2 and 2,1 giving us the factor of two in the double sum). If the random variables are statistically independent, that is $\rho = 0$, then the double sum goes to zero and we are left with:

$$\sigma_Z^2 = \sum_{i=1}^{n} a_i^2 \sigma_{Q_i}^2 \qquad equation\ 5.7$$

Notation Clarity

Depending on how i's and j's are treated will dictate how the variance can be written. The following are all equivalent notations resulting in the exact same solution for the squared second moment. The first is the most concise while the third is the most explicit.

$$\sigma^2 = \sum_{i=1}^{n} \sum_{j=1}^{n} a_i a_j \rho_{ij} \sigma_i \sigma_j \qquad \textit{equation } 5.8a$$

$$\sigma^2 = \sum_{i=1}^{n} a_i^2 \sigma_i^2 + \sum_{i,j=1}^{n} \sum_{i \neq j}^{n} a_i a_j \rho_{ij} \sigma_i \sigma_j \qquad \textit{equation } 5.8b$$

$$\sigma^2 = \sum_{i=1}^{n} a_i^2 \sigma_i^2 + 2 \sum_{i=1}^{n} \sum_{j=i+1}^{n} a_i a_j \rho_{ij} \sigma_i \sigma_j \qquad \textit{equation } 5.8c$$

The second special case is where the function is a product or quotient of lognormally distributed random variables. This is because when we take the natural log of a product we have the sum, or when we take the natural log of a quotient we have the difference. And the sum or difference will be of a log transformed normal distribution.

$Z = X \cdot Y$ becomes $lnZ = lnX + lnY$ when we log both sides.

$Z = X/Y$ becomes $lnZ = lnX - lnY$ when we log both sides.

Where: lnX is $N(\mu_X, \sigma_X)$ and X is $LN(\lambda_X, \xi_X)$
lnY is $N(\mu_Y, \sigma_Y)$ and Y is $LN(\lambda_Y, \xi_Y)$

Therefore: lnZ is $N(\mu_Z, \sigma_Z)$ and Z is $LN(\lambda_Z, \xi_Z)$

We are now back to a sum or difference of normally distributed random variables. When we have the product of X and Y which are lognormally distributed random variables the first and second moments are:

$$\lambda_Z = \lambda_X + \lambda_Y \qquad equation\ 5.9$$

$$\xi_Z^2 = \xi_X^2 + \xi_Y^2 + 2\rho\xi_X\xi_Y \qquad equation\ 5.10$$

And when the random variables are statistically independent:

$$\xi_Z^2 = \xi_X^2 + \xi_Y^2 \qquad equation\ 5.11$$

This can be similarly expanded for any number of random variables with coefficients in the same manner as with normally distributed random variables, using the same equivalent notation as presented in Notation Clarity.

Example: Waste Treatment Facility

The annual operating cost function for waste treatment plants can be written:

$$C = \frac{WF}{\sqrt{E}}$$

Where W is waste, F is the cost factor, and E is efficiency on an annualized basis. All these parameters are treated as lognormal random variables in this problem, $LN(\lambda, \xi)$.

	median value	coefficient of variation
W	2000 metric tons/yr	20%
F	\$20/metric ton	15%
E	1.6	12.5%

With the given information, what is the probability that the cost will be greater than \$35,000 and the waste contractor must make financial contingency plans for the year?

Because the function is a product/quotient and the random variables are lognormal we can use an exact solution. We first solve for the moments of C. Taking the natural log of both sides of the function:

$$lnC = lnW + lnF - \frac{1}{2}lnE$$

$$\lambda_C = \lambda_W + \lambda_F - \frac{1}{2}\lambda_E$$

$$= ln2000 + ln20 - \frac{1}{2} ln1.6 = 10.36$$

$$\xi_C^2 = \xi_W^2 + \xi_F^2 + \left(-\frac{1}{2}\right)^2 \xi_E^2$$

$$= (0.20)^2 + (0.15)^2 + \left(-\frac{1}{2}\right)^2 (0.125)^2 = 0.067$$

$$\xi_C = \sqrt{0.067} = 0.259$$

Now that we have the moment of C we can evaluate the probability of exceedance.

$$P(C > 35{,}000) = 1 - P(C \leq 35{,}000) \quad compliment$$

$$= 1 - \Phi\left(\frac{ln35{,}000 - \lambda_C}{\xi_C}\right) \quad as\ C\ is\ LN(\lambda_C, \xi_C)$$

$$= 1 - \Phi(0.398) \cong 1 - 0.655 = \mathbf{0.345}$$

There is roughly a 35% chance that the annual cost will exceed $35,000. The waste contractor would most likely prepare for a cost overrun given these odds. If the decision was close the contractor may want to spend more time on the probability analysis by investigating the distributions of the independent variables.

In summary, for the special cases of sum/difference of normally distributed random variables, and product/quotient of lognormally distributed random variables we can solve for the full distribution of the resultant. And in the situation where we have a simple one-to-one function with a single root we can also solve for the full distribution of the resultant. These are the two situations where we have exact solutions.

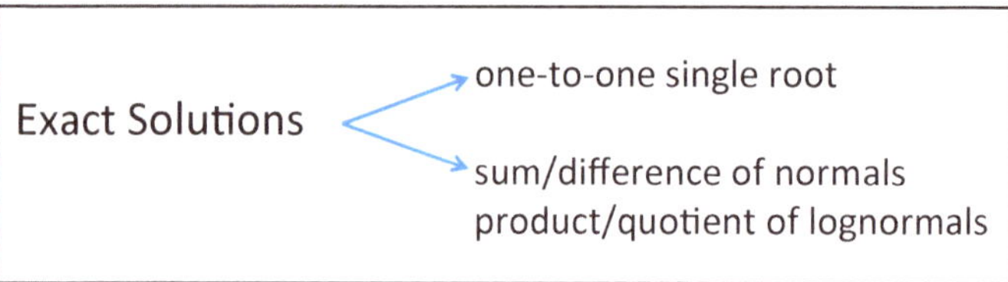

If, however, we have a function or random variables that don't fit these cases (which is typical in Civil Engineering) then we must resort to other solution techniques.

Approximate Solutions

There are two common approximate solutions. The first invokes the central limit theorem (CLT), the second uses a Taylor series expansion.

Central Limit Theorem Approximation

The central limit theorem states that the sum of a large number of individual random components, none of which are significantly more dominant than the others, tends towards a Gaussian distribution as the number of components increases. This holds true regardless of the underlying distribution of the individual components. This would then give us similar results as the sum/difference of normally distributed random variables:

$$Z = \sum_{i=1}^{n} a_i Q_i \qquad equation\ 5.12$$

$$\mu_Z \approx \sum_{i-1}^{n} a_i \mu_{Q_i} \qquad equation\ 5.13$$

$$\sigma_Z^2 \approx \sum_{i=1}^{n} a_i^2 \sigma_{Q_i}^2 + \sum_{i,j=1}^{n} \sum_{i \neq j}^{n} a_i a_j \rho_{Q_i Q_j} \sigma_{Q_i} \sigma_{Q_j} \qquad equation\ 5.14$$

As $i \rightarrow \infty$ then Z approaches the normal distribution, $N(\mu_Z, \sigma_Z)$, even if the $Q's$ are not normally distrusted. The drawback with this approximation is that we don't know how large i needs to be to give a good approximation, and it requires validation in most cases.

Example: Linear Equation

The equation below is a sum of random variables (some with negative coefficients) and the specified correlation.

$$Y = V - T + G - Q \quad where \quad \rho_{VT} = 0.5$$

Invoking the Central Limit Theorem to approximate the moments of the dependent variable:

$$\mu_Y \approx \mu_V + (-1)\mu_T + \mu_G + (-1)\mu_Q$$

$$\sigma_Y^2 \approx \sigma_V^2 + (-1)^2\sigma_T^2 + \sigma_G^2 + (-1)^2\sigma_Q^2 + 2(1)(-1)0.5\sigma_V\sigma_T$$

The assumption with this solution is that the dependent variable is Gaussian zwith the above moments.

First Order Second Moment

A more verifiable solution is what is called the Mean Value First Order Second Moment or more simply the First Order Second Moment (FOSM) approach. This solution works for any type of function and with variables of any distribution. The drawback is that we can only solve for the first and second moments, and not the full distribution.

With this method we can find the moments of any functional form or mathematical equation $g(X)$ of a random variable X. If we say that:

$$Y = g(X) \qquad equation\ 5.15$$

Then the first and second moments would be:

$$E(Y) = \int_{-\infty}^{\infty} g(X) \cdot f(x)dx \qquad equation\ 5.16$$

$$Var(Y) = \int_{-\infty}^{\infty} [g(x) - \mu_x]^2 \cdot f(x)dx \qquad equation\ 5.17$$

To solve for the mean and variance of X we need the PDF $f(X)$. But in many cases the PDF is not available or assuming a PDF introduces too much epistemic uncertainty, which lends to an approximate solution for the mean and the variance. To approximate the moments we expand the function in a Taylor series about the mean value of X:

$$g(X) = g(\mu_X) + (X - \mu_X)\frac{dg}{dX} + \frac{1}{2}(X - \mu_X)^2\frac{d^2g}{dX^2} + \cdots \qquad equation\ 5.18$$

Where the derivatives are also evaluated at the mean value of X. It is common to truncate the series above the linear terms thereby giving a first order approximate of the moments:

$$g(X) \approx g(\mu_X) + (X - \mu_X)\frac{dg}{dX} \qquad equation\ 5.19$$

$$E(Y) = \mu_Y \approx g(\mu_x) \qquad equation\ 5.20$$

$$Var(Y) = \sigma_Y^2 \approx \sigma_X^2 \left(\frac{dg}{dX}\right)^2 \qquad equation\ 5.21$$

The first order approximate will provide reasonable accuracy in situations where the function in not highly non-linear and in situations where the variance is not large. If more accuracy is warranted a second order approximation can be carried out using higher order terms in the Taylor series expansion. Here we reevaluate the mean using a second order approximate:

$$E(Y) \approx g(\mu_x) + \frac{1}{2}\sigma_X^2 \frac{d^2g}{dX^2} \qquad equation\ 5.22$$

In most Civil Engineering problems a first-order (FOSM) estimate is sufficient for quantifying the mean and standard deviation. We can expand this discussion to include multiple random variables:

$$Y = g(X_1, X_2, \dots, X_n) \qquad equation\ 5.23$$

$$E(Y) = \mu_Y \approx g\left(\mu_{X_1,}\mu_{X_2}, \dots, \mu_{X_n}\right) \qquad equation\ 5.24$$

$$Var(Y) = \sigma_Y^2 \approx \sum_{i=1}^{n} \sigma_{X_i}^2 \left(\frac{\partial g}{\partial X_i}\right)^2 + \sum_{i,j=1}^{n} \sum_{i \neq j}^{n} \rho_{X_i X_j} \sigma_{X_i} \sigma_{X_j} \frac{\partial g}{\partial X_i}\frac{\partial g}{\partial X_j} \qquad equation\ 5.25$$

When the random variables, $X's$, are statistically independent, that is $\rho = 0$, then the double sum goes to zero and we are left with:

$$Var(Y) = \sigma_Y^2 \approx \sum_{i=1}^{n} \sigma_{X_i}^2 \left(\frac{\partial g}{\partial X_i}\right)^2 \qquad equation\ 5.26$$

If a higher order estimate of the mean for multiple random variables is desired, the second-order Taylor series expansion takes the form:

$$E(Y) = \mu_Y \approx g(\mu_{X_1,}\mu_{X_2}, \dots, \mu_{X_n}) + \frac{1}{2}\sum_{i=1}^{n}\sum_{j=1}^{n} \rho_{ij}\sigma_{X_i}\sigma_{X_j}\left(\frac{\partial^2 g}{\partial X_i \partial X_j}\right) \qquad equation\ 5.27$$

If the independent variables are uncorrelated then ***equation 5.27*** reduces to:

$$E(Y) = \mu_Y \approx g(\mu_{X_1,}\mu_{X_2}, \dots, \mu_{X_n}) + \frac{1}{2}\sum_{i=1}^{n} \sigma_{X_i}^2\left(\frac{\partial^2 g}{\partial X_i^2}\right) \qquad equation\ 5.28$$

Example: Equation of a Line

In this example we will evaluate the equation of a line where the parameters are all random variables with their respective means and standard deviations. This would be the case if we performed a linear regression while keeping track of the uncertainty in the input data and the slope and intercept terms. Here we will assume that the random variables are statistically independent but that would obviously not be the case for the slope and intercept terms. Using FOSM to approximate the moments of the dependent variable:

$$y = m \cdot x + b$$

$$\mu_y \approx \mu_m \cdot \mu_x + \mu_b$$

$$\sigma_y^2 \approx \sigma_m^2\left(\frac{\partial y}{\partial m}\right)^2 + \sigma_x^2\left(\frac{\partial y}{\partial x}\right)^2 + \sigma_b^2\left(\frac{\partial y}{\partial b}\right)^2$$

$$where \quad \frac{\partial y}{\partial m} = x, \quad \frac{\partial y}{\partial x} = m, \quad \frac{\partial y}{\partial b} = 1$$

$$\sigma_y^2 \approx \sigma_m^2(x)^2 + \sigma_x^2(m)^2 + \sigma_b^2$$

$$\approx \sigma_m^2 \cdot \mu_x^2 + \sigma_x^2 \cdot \mu_m^2 + \sigma_b^2 \quad about\ the\ mean$$

One benefit of performing FOSM is that it provides a sensitivity analysis on the function and how the uncertainty of the variables individually influence the resultant. If we look at the variance of a function of random variables we see that the uncertainty from the input parameter propagates to the resultant as the product of the variance and the squared partial derivative of the function with respect to the input parameter. The variance of the input parameter is weighted according to how it participates in the mathematics of the function. When divided by the total resultant uncertainty this provides the percent relative contribution to variance (*RCV%*) from each random variable:

$$RCV\% = \sigma_{X_i}^2 \left(\frac{\partial g}{\partial X_i}\right)^2 \Big/ \sigma_Y^2 \qquad equation\ 5.29$$

Example: Manning's Equation

Manning's equation is an empirical equation that is used to determine the velocity (V) in meters per second of uniform flow in an open channel:

$$V = \frac{R^{2/3}\, S^{1/2}}{n}$$

Where:
R = hydraulic radius (m)
S = slope of the energy line (%)
n = empirical roughness coefficient of the channel

If the channel is an open concrete rectangular section the mean values and coefficients of variation of the variables are:

Variable	μ	δ
R	2 m	0.05
S	1 %	0.10
n	0.013	0.10

If we assume the variables are statistically independent, then the FOSM solution of the mean and variance is:

$$\mu_V \approx \frac{\mu_R^{2/3}\ \mu_S^{1/2}}{\mu_n} = \frac{(2)^{2/3}(1)^{1/2}}{0.013} = 122.108\ mps$$

$$\sigma_V^2 \approx \sigma_R^2 \left(\frac{2}{3} \frac{\mu_R^{-1/3} \mu_S^{1/2}}{\mu_n}\right)^2 + \sigma_S^2 \left(\frac{1}{2} \frac{\mu_R^{2/3} \mu_S^{-1/2}}{\mu_n}\right)^2 + \sigma_n^2 \left(-\frac{\mu_R^{2/3} \mu_S^{1/2}}{\mu_n^2}\right)^2$$

$$= (2 \cdot 0.05)^2 \left(\frac{2}{3} \frac{(2)^{-2/3} (1)^{1/2}}{0.013}\right)^2 + (1 \cdot 0.01)^2 \left(\frac{1}{2} \frac{(2)^{2/3} (1)^{-1/2}}{0.013}\right)^2 + (0.013 \cdot 0.10)^2 \left(-\frac{(2)^{2/3} (1)^{1/2}}{(0.013)^2}\right)^2$$

$$= (0.01 \cdot 2485.052) + (0.01 \cdot 3727.576) + (0.0000169 \cdot 88226676.230)$$

$$= 24.851 + 37.276 + 149.103 = 211.229$$

$$\sigma \approx \sqrt{211.23} = 14.534$$

For this problem the relative contribution to the total variance (*RCV%*) is: ***R*** of 11.76%, ***S*** of 17.65%, and ***n*** of 70.59%. It can be seen that the uncertainty in the input roughness coefficient, *n*, has the largest impact on the total uncertainty of the resultant velocity, *V*. This relative uncertainty impact is commonly expressed through a sensitivity study. Each variable within a function is varied about its mean by plus/minus one standard deviation (±sigma), and the effect on the dependent variable is then shown as a "tornado" plot. Visually we see how the uncertainty of the roughness coefficient, *n*, has the most influence on the resultant velocity, *V*, in this problem.

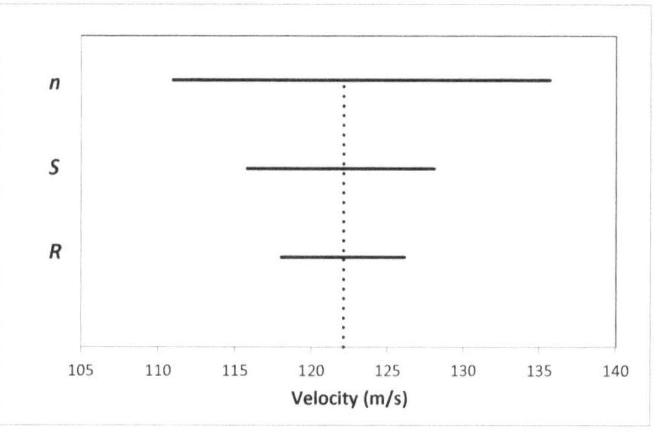

To check the accuracy of the solution we can evaluate the second-order approximation of the mean velocity, *V*, following *equation 5.28*:

$$\mu_V \approx \mu_V + \frac{1}{2}\left[\sigma_R^2\left(-\frac{2}{9}\frac{\mu_R^{-4/3}\,\mu_S^{1/2}}{\mu_n}\right) + \sigma_S^2\left(-\frac{1}{4}\frac{\mu_R^{2/3}\,\mu_S^{-3/2}}{\mu_n}\right) + \sigma_n^2\left(2\,\frac{\mu_R^{2/3}\,\mu_S^{1/2}}{\mu_n^3}\right)\right]$$

$$= 122.11 + \frac{1}{2}\left[(2\cdot 0.05)^2\left(-\frac{2}{9}\frac{(2)^{-4/3}\,(1)^{1/2}}{0.013}\right) + (1\cdot 0.01)^2\left(-\frac{1}{4}\frac{(2)^{2/3}\,(1)^{-3/2}}{0.013}\right) + (0.013\cdot 0.10)^2\left(2\,\frac{(2)^{2/3}\,(1)^{1/2}}{(0.013)^3}\right)\right]$$

$$= 122.108 + \frac{1}{2}[-0.068 - 0.305 + 2.442] = 123.142$$

The first-order estimate is less than 1% lower than the second order estimate, providing confidence in the approximate solution of this problem.

In summary, for situations that don't lend to an exact solution we can approximate the solution using the two methods; CLT or FOSM. The central limit theorem approach is limited to a sum/difference function and the resultant distribution is assumed normal, but not verified.

The first order second moment approach works for any type of mathematical function with random variables of any distribution. FOSM is an extremely versatile approach and works when the moments are sufficient information to proceed with the problem. For most Civil Engineering problems this is adequate, therefore FOSM is the "workhorse" of propagating uncertainty.

Approximate Solutions
- → CLT: sum/difference as N→∞
- → FOSM: any mathematical function with random variables of any distribution

A means of confirming the approximate solutions and providing confidence in the answer is a computational solution as discussed in the next section.

Monte Carlo Simulations

Monte Carlo (MC) simulations are a "brute force" computation approach to solve probabilistic problems. MC simulations are used in a wide variety of fields, and not restricted to probability problems, giving rise to a vast depth of literature on different sampling techniques, computational efficiency, and numerical algorithms (e.g., Ripley, 1987). For our purposes we will be using MC simulations for functions of random variables to compliment the exact or approximate solutions.

The term Monte Carlo comes from the casino in Monaco and is named thus because the method relies on a random simulation of outcomes similar to the process of gambling. In order to solve the simple sum problem that was posed at the beginning of the chapter with a Monte Carlo simulation we;

- randomly simulate a realization from the discrete distribution of X,
- randomly simulate a realization from the continuous distribution of Y,
- add the two realizations together as this particular mathematical function calls for, which produces a realization of Z,
- then repeat this process many many times to characterize the full distribution of Z.

We are in essence discretizing the entire problem, generating histograms of the input parameter distributions, and carrying out the function's mathematics a very large number of times until we have fully developed the resultant distribution.

With rapid computational speeds and pre-programmed probability distributions we can accomplish 10,000 to 1,000,000 simulations in fractions of a second. The resultant distribution is an approximation of the answer but as the number of simulations becomes large the line between approximate and exact becomes blurred, dictated by machine precision or machine epsilon.

Generating random realizations can be accomplished using any number of computational programs (e.g., MATLAB, MathCad, Excel). In this text we will be focusing on MATLAB because of its ease and capabilities in simulating numbers, but the discussions pertain to any similar computational program.

There is an entire body of literature that discusses how to generate a random number based on a fixed computer algorithm (e.g., Knuth, 1997), which is an inherently contradictory problem. For our purposes we assume that the random number generated is random enough for our calculations.

The best way to discuss Monte Carlo simulations methods is through example.

Example: Sum of Lognormals

We need to analyze the sum function:

$$S = X_1 + X_2 + X_3$$

If the input parameters were normally distributed then we could use an exact solution, but in this problem we are given that the input parameters are lognormally distributed, X_i are $LN(\lambda_i, \xi_i)$. The sum of lognormal variables does not produce a lognormal resultant.

We can use FOSM to approximate the moments of S, given the moments of the X_i. We can also use Monte Carlo simulations, and then compare the results of the two methods. The information given for the X_i:

	X_1	X_2	X_3
μ	500	600	700
δ	0.50	0.60	0.70

Since no correlation coefficient is given we assume that the X_i are statistically independent.

FOSM gives (which is the same answer arrived at using CLT approximation):

$$\mu_S \approx \mu_{X_1} + \mu_{X_2} + \mu_{X_3} = 500 + 600 + 700 = 1800$$

$$\sigma_S^2 \approx \sigma_{X_1}^2 \left(\frac{\partial g}{\partial X_1}\right)^2 + \sigma_{X_2}^2 \left(\frac{\partial g}{\partial X_2}\right)^2 + \sigma_{X_3}^2 \left(\frac{\partial g}{\partial X_3}\right)^2 = \sigma_{X_1}^2 \cdot 1 + \sigma_{X_2}^2 \cdot 1 + \sigma_{X_3}^2 .1$$
$$\approx (500 \cdot 0.50)^2 + (600 \cdot 0.60)^2 + (700 \cdot 0.7)^2 = 432{,}200$$

$$\sigma_S \approx \sqrt{432{,}200} = 657$$

MC simulations are performed using MATLAB with the following m-file to carry out the simulations:

```
%%%%%%%%%%%%%%%%%%%%%%%%%%%%%%%%%%%%%%%%%%%%%%%%%%%%%%%%
%Using MC simulations on S=X1+X2+X3 where X's are LN

%first we need to calculate the moments of the
%lognormal distributions from their given means and
%coefficients of variations
```

```
xsi_1=sqrt(log(1+0.5^2));
lambda_1=log(500)-0.5*(xsi_1)^2;

xsi_2=sqrt(log(1+0.6^2));
lambda_2=log(600)-0.5*(xsi_2)^2;

xsi_3=sqrt(log(1+0.7^2));
lambda_3=log(700)-0.5*(xsi_3)^2;

%next we simulation the 3 lognormal random variables
n=10000; %number of simulations

x1=lognrnd(lambda_1,xsi_1,n,1);
x2=lognrnd(lambda_2,xsi_2,n,1);
x3=lognrnd(lambda_3,xsi_3,n,1);

%then we perform the calculation sequentially
s=x1+x2+x3;

%Results
hist(s,50) %the discrete results of n simulations
x_bar=mean(s) %sample mean value of n simulations
s=std(s)  %sample standard deviation of n simulations
%%%%%%%%%%%%%%%%%%%%%%%%%%%%%%%%%%%%%%%%%%%%%%%%%%%%%
```

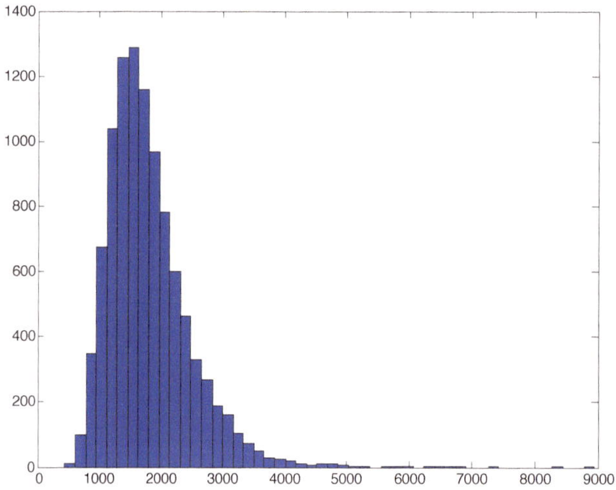


```
x_bar=1.7942e+03
s=660.2819
```

The histogram shows a skewed distribution but if we were to test it we would find that it does not exactly follow the lognormal distribution. (It is beyond the scope of this text but the Gamma distribution is often useful for modeling a skewed distribution like the results here.)

Results will vary slightly for each run of the subroutine, even with the same number of simulations, because the realizations are randomly generated. But as we approach a large number of simulations the results will converge. This is a way of determining the accuracy of the simulations. The figure below shows the change in the sample mean as the number of simulations are increased.

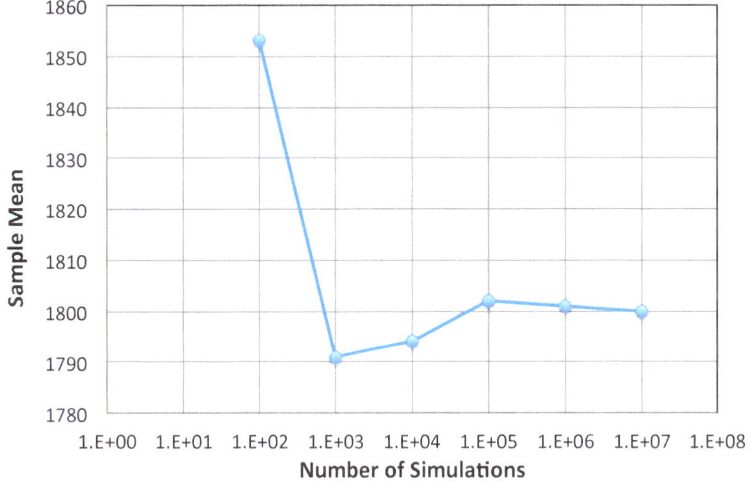

We can see that in this problem the answer converges around 100,000 (1.E+05) to 1,000,000 (1.E+06) simulations. The computational cost of increased simulations is measured in runtime, where it took 0.050 seconds to generate 100,000 simulations compared to 0.284 seconds for 1,000,000 simulations. Both runs are nominal computationally, but if we increase the simulations in this problem up to 10,000,000 (1.E+07) the run time is 2.370 seconds, a noticeable delay for the user. Of course the run time will increase with more complex problems/calculations, and more efficient programming can help decrease runtime.

If we compare the MC simulations with the FOSM approximations we find there is good agreement which provides confidence in our answer. This also demonstrates that for linear functions with input parameters that have relatively symmetric distributions, FOSM can often give accurate results.

FOSM	MC (1.E+07) simulations
$\mu = 1800$ $\sigma = 657$	x_bar=1800 s=657

It is common to use both an approximate and computational method to provide confidence in the answer and to tease out if there is any nonlinearity or other aspects contributing to the results.

Example: Product/Quotient of Normals

In this problem we will be evaluating the function:

$$P = \frac{X_1 X_3}{X_2}$$

The random variables in this function are all normally distributed with their respective moments, X_i are $N(\mu_i, \sigma_i)$. Note, if these were all lognormally distributed then we could solve using an exact solution, the product/difference of lognormals. Since they are normally distributed we must resort to an approximate solution and/or MC simulations. The information given for the X_i are:

	X_1	X_2	X_3
μ	500	600	700
σ	75	120	210

No correlation coefficient was given so, as before, we will assume that the variables are statistically independent.

The FOSM approximate of this is:

$$\mu_P \approx \frac{\mu_{X_1}\mu_{X_3}}{\mu_{X_2}} = 583$$

$$\sigma_P^2 \approx \sigma_{X_1}^2 \left(\frac{X_3}{X_2}\right)^2 + \sigma_{X_2}^2 \left(-\frac{X_1 X_3}{{X_2}^2}\right)^2 + \sigma_{X_3}^2 \left(\frac{X_1}{X_2}\right)^2$$

$$\approx 75^2 \left(\frac{700}{600}\right)^2 + 120^2 \left(-\frac{500 \cdot 700}{600^2}\right)^2 + 210^2 \left(\frac{500}{600}\right)^2 = 51892.36$$

$$\sigma_P \approx 227.8$$

If we now wanted to know the probability that P would exceed 700, then we could approximate the probability using the standard normal distributions.

$$P(P \geq 700) = 1 - P(P \leq 700) \approx 1 - \Phi\left(\frac{700 - 500}{227.8}\right) = 1 - 0.8105$$
$$\approx \mathbf{0.1894} \text{ or roughly } \mathbf{19\%}$$

Solving the problem using MC simulations the m-file looks like:

```
%%%%%%%%%%%%%%%%%%%%%%%%%%%%%%%%%%%%%%%%%%%%%%%%%%%%%%%
%Using MC simulations to solve P=(X1*X3)/X2 where X's
%are normal

%first we declare the mean's and standard deviations
mu_1=500;
sigma_1=75;

mu_2=600;
sigma_2=120;

mu_3=700;
sigma_3=210;

%next we simulate the normal random variables
n=10000;

x1=normrnd(mu_1,sigma_1,n,1);
x2=normrnd(mu_2,sigma_2,n,1);
x3=normrnd(mu_3,sigma_3,n,1);

%then we perform the calculation sequentially.
%note: the dots in the equation specifies that linear,
%not matrix, mathematics be carried out
p=(x1.*x3)./x2;

%Results
hist(p,50)
x_bar=mean(p)
s=std(p)
prob=(sum(p>700))/n

%note: because these are simulations we can sum all the
%realizations that exceed the threshold of 700 and
%divide by the total number of simulations n to
%estimate the probability
%%%%%%%%%%%%%%%%%%%%%%%%%%%%%%%%%%%%%%%%%%%%%%%%%%%%%%%
```

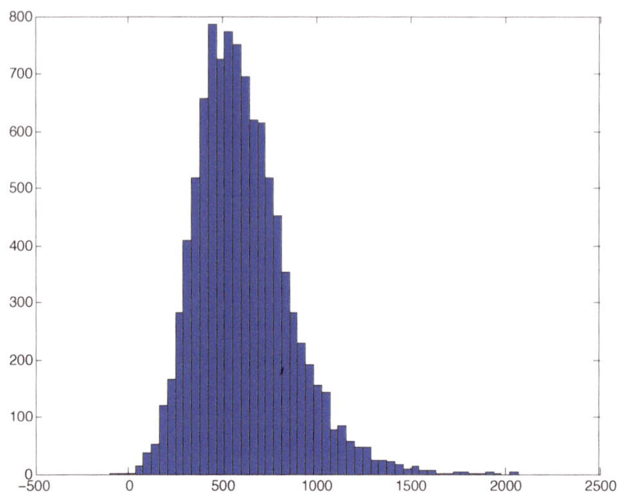

Results for (1.E+07) simulations
x_bar =610.2443
s=261.0716
prob =0.3060

In this problem it is interesting to observe that the resultant distribution is neither normal nor lognormal but something inbetween. Now comparing the results of FOSM and MC simulations we find that in this case, with a nonlinear equation, FOSM underestimates the mean, standard deviation, and probability of exceedance. In this particular problem the MC simulations more accurately model the interaction of the input parameter distributions and the functional form of the equation.

FOSM	MC (1.E+07) simulations
$\mu = 583$	x_bar=610.2
$\sigma = 227.8$	s=261.1
$P(P \geq 700) \approx 19\%$	prob =0.3060

As has been demonstrated, MC simulations for statistically independent random variables are relatively straight forward. When random variables are correlated we need to generate joint distributions to account for that correlation.

The **joint distribution** of two variables can be written with the multiplication rule, here using CDF's:

$$F(xy) = F(y|x)F(x) \quad \textit{equation } 5.31$$

If we have the marginal distribution of X and the conditional distribution of Y given X, the we can solve for the joint distribution. Computationally the joint distribution can be found using the algorithm presented in Ang and Tang (1984) for two random variables:

1) Generate two uniformly distributed vectors, u_1 and u_2, between 0 and 1.
2) Random variable X is generated as the inverse CDF of the first uniformly distributed random vector: $X = F^{-1}(u_1)$.
3) Random variable Y is generated as the inverse CDF of the second uniformly distributed random vector conditioned on X: $Y = F^{-1}(u_2|x)$

This algorithm applies to any joint distribution and many inverse CDF's are pre-programmed into computational software (e.g., normal, lognormal, gamma, beta). For joint distributions that are not pre-programmed the inverse CDF must be solved symbolically. This can be accomplished using MATLAB but lies out side the scope of this text and readers are referred to Ang and Tang (2007) for more details. For **joint normal** the distributions of Gaussian X and $Y|X$ are:

$$X = \Phi^{-1}(u_1)\sigma_X + \mu_X \quad \textit{equation } 5.32$$

$$Y = \Phi^{-1}(u_2)\sigma_Y\sqrt{1-\rho^2} + \left(\mu_Y + \rho\left(\frac{\sigma_Y}{\sigma_X}\right)(x - \mu_X)\right) \quad \textit{equation } 5.33$$

The following example illustrates this calculation in MATLAB.

Example: Joint Normals

In this problem we are evaluating the function:

$SQ = X - Y$

Where X and Y are assumed jointly normal and negatively correlated. We can solve this using an exact solution and then verify using MC simulations. The statistics provided for the input parameters are:

	X	*Y*
$\bar{x}$	500	600
s	250	360
ρ	-0.33	

An exact solution produces:

$$\mu_{SQ} = \mu_X + \mu_Y = 500 + (-600) = -100$$

$$\begin{aligned}\sigma_{SQ}^2 &= a_X^2\sigma_X^2 + a_Y^2\sigma_Y^2 + 2\rho a_X a_Y \sigma_X \sigma_Y \\ &= (1)^2 250^2 + (-1)^2 360^2 + 2(-0.33)1(-1)250(360) \\ &= 62500 + 129600 + 59400 = 251500\end{aligned}$$

$$\sigma_{SQ} = \sqrt{251500} = 501.5$$

If we are interested in the probability that SQ will exceed 500, following the exact solution we find:

$$\begin{aligned}P(SQ > 500) = 1 - P(SQ \leq 500) = 1 - \mathrm{F}(500) \\ = 1 - \Phi\left(\frac{500 - (-100)}{501.5}\right) \approx 1 - 0.8842 = 0.1158\end{aligned}$$

The solution using MC simulations uses the following script:

```
%%%%%%%%%%%%%%%%%%%%%%%%%%%%%%%%%%%%%%%%%%%%%%%%%%%%%%%
%Using MC simulations to solve SQ=X-Y
%where X and Y are jointly normal and correlated

%first we declare the knowns
mu_x=500;
sigma_x=250;

mu_y=600;
sigma_y=360;

rho=-0.33;

%next we simulate the uniformly distributed vectors
n=1000000;
u1=unifrnd(0,1,n,1);
u2=unifrnd(0,1,n,1);
```

```
%then we generate the normal distribution of X
x=(norminv(u1).*sigma_x)+mu_x;

%and the conditional distribution of Y given X that
%includes correlation
y=(norminv(u2).*sigma_y.*sqrt(1-
   rho^2))+(mu_y+rho*(sigma_y/sigma_x).*(x-mu_x));

%now evaluating the function
sq=x-y;

%Results
hist(sq,100);
x_bar=mean(sq)
s=std(sq)
prob=(sum(sq>500)/n)
%%%%%%%%%%%%%%%%%%%%%%%%%%%%%%%%%%%%%%%%%%%%%%%%%%%%%%%%%%%%%%%%%%%%%%%%%%%%%%%%%%%%%%%%%%%%%%%%%%%%%%%%%%%%%%%%%%
```

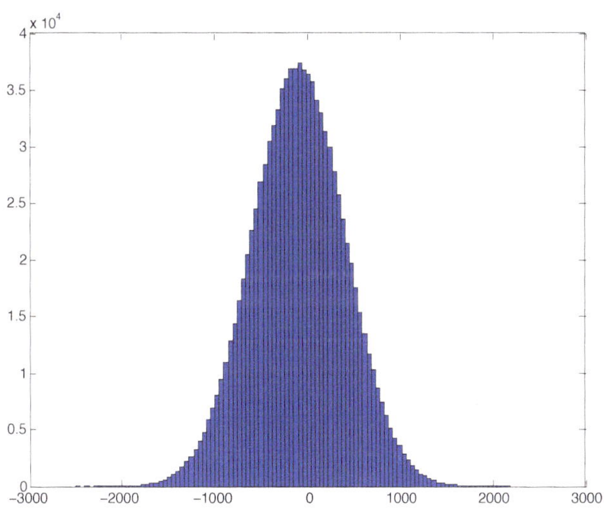

x_bar =-99.9253

s =501.6675

prob =0.1156

The histogram shows a normally distributed resultant as would be expected when we have the difference of normal distributions. Comparing the numeric results, we see that MC simulations provides a very close answer to the exact solution. This provides a check on our work and confidence in moving forward with the probability of exceedence for decision purposes.

Exact	MC simulations
$\mu = -100$	x_bar =-99.9253
$\sigma = 501.5$	s=501.6675
$P(SQ > 500) \approx 11.58\%$	prob =0.1156

The above code can be written using the multivariate function, which eliminates the need for explicitly generating the uniformly distributed random vectors.

```
%%%%%%%%%%%%%%%%%%%%%%%%%%%%%%%%%%%%%%%%%%%%%%%%%%%%%%%%%%%%%
%Using MC simulations to solve SQ=X-Y where X and Y are
%jointly normal with prescribed correlation coefficient

%declare the means and standard deviations
mu_x=500;
sigma_x=250;

mu_y=600;
sigma_y=360;

rho=-0.33;

%setting up the means vector and covariance matrix
mu_vector=[mu_x mu_y];
%see Appendix C for discussion of the covariance matrix
cov_matrix=[sigma_x^2           rho*sigma_x*sigma_y ;
            rho*sigma_x*sigma_y           sigma_y^2];

%simulate the joint normal distribution
n=1000000;
xy=mvnrnd(mu_vector,cov_matrix,n);

%evaluate the function
sq=xy(:,1)-xy(:,2);

%Results
hist(sq,100);
x_bar=mean(sq)
s=std(sq)
prob=(sum(sq>500)/n)
%%%%%%%%%%%%%%%%%%%%%%%%%%%%%%%%%%%%%%%%%%%%%%%%%%%%%%%%%%%%%
```

The results using the multivariate function (mvnrnd) provides nearly identical results as before.

Exact	MC simulations
$\mu = -100$	x_bar =-99.9746
$\sigma = 501.5$	stdev =502.0354
$P(SQ > 500) \approx 11.58\%$	prob =0.1159

For jointly lognormal parameters the algorithm using uniformly distributed random variables can be similarly used to solve problems. MATLAB does not currently support a multivariate lognormal function at the time of writing, however one is available in the user forum which performs the desired correlated lognormal simulation[§].

In summary, the discussion of computational methods, in using Monte Carlo simulations we can generate sequential realizations of the random variables of interest, and plug those realizations into the function we are evaluating. This provides a simulation as accurate as needed to solve any function of random variables. The normal and lognormal distributions are easy to simulate because these distributions are pre-programmed into computational software. Monte Carlo simulations provide a means of checking approximate solutions and can often be used to tackle more complex functions with difficult partial derivatives that can render FOSM intractable.

Computational Solutions → Monte Carlo (MC) simulations: uncorrelated or correlated

Chapter Summary

- Solving a **function of random variables** is the core material of this text. In almost all Civil Engineering problems we are using some mathematical equation made up of parameters that contain uncertainty. Translating or propagating this uncertainty from the

---

[§]http://www.mathworks.com/matlabcentral/fileexchange/6426-multivariate-lognormal-simulation-with-correlation accessed 2/4/2013

input parameters to the resultant gives us a measure of how accurate the solution is.

- Exact solutions exist for one-to-one single root functions, for a function that is the sum/difference of normally distributed random variables, and for a function that is the produce/difference of lognormally distributed random variables.
- For situations that don't fit the exact solutions we can use **approximation methods** and/or **Monte Carlo (MC) simulations**.
- Two approximate methods presented are the **Central Limit Theorem** (CLT) and the **First Order Second Moment** (FOSM) methods.
- CLT assumes the resultant is normal given a large number of input parameters in a sum/difference function.
- FOSM uses a **Taylor series** expansion about the mean to estimate the moments of a function given the moments of the input parameters. FOSM works on any mathematical function with input parameters having any distribution.
- MC simulations uses many many realizations to estimate the resulting distribution.
- MC simulations methods are often used in conjunction with other methods to validate the answer.

6 COMPONENT RELIABILITY ANALYSIS

Reliability at the component level is the probabilistic relationship between load and resistance, or stress and strain, or demand and capacity; these paired terms are all analogous. Reliability is commonly expressed using the reliability index, β, which can in turn be related to the probability of failure, p_f.

Figure 6.1 shows a deterministic view of an engineering problem. Load (Q) and Resistance (R) are shown on a number line with respect to each other. Whenever the load is less than the resistance then the design is considered safe. Using a factor of safety formulation where $FS = R/Q$, a "no failure" state is where FS>1 when Q is less than R. Using a margin of safety formulation where $M = R - Q$, a "no failure" state is where M>0 when Q is less than R.

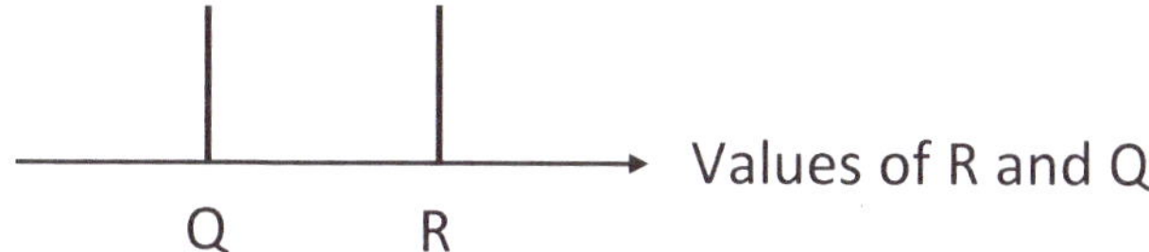

Figure 6.1 Deterministic view of load (Q) versus resistance (R).

When we include the uncertainty of the load and resistance in the analysis, as Figure 6.2 shows there may be a region where failure can occur. Using factor of safety formulation, $P(FS) = P(R)/P(Q)$, or margin of safety formulation, $P(M) = P(R) - P(Q)$, and the same mean values as in the deterministic view, if there is sufficient uncertainty in the load and/or resistance then there is a probability of failure.

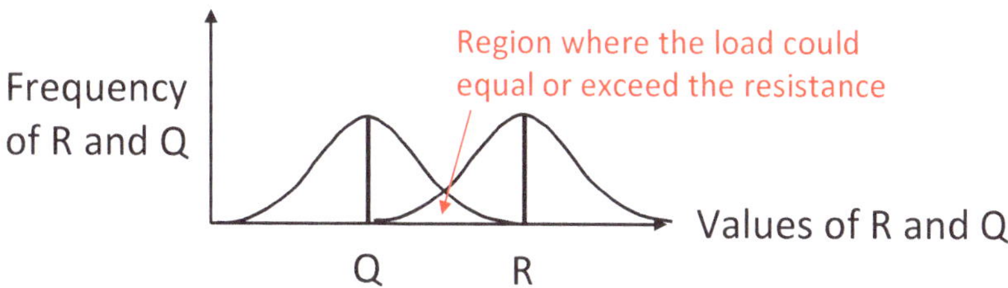

Figure 6.2 Probabilistic view of load (Q) versus resistance (R).

Component Reliability Formulation

The fundamental presentation of component reliability can be accomplished using the margin of safety formulation and assuming that the load and resistance are jointly normal.

$$M = R - Q \; where \; R \; and \; Q \; are \; jointly \; normal \qquad equation\ 6.1$$

$$\mu_M = \mu_R - \mu_Q \qquad equation\ 6.2$$

$$\sigma_M^2 = \sigma_R^2 + \sigma_Q^2 - 2\rho_{RQ}\sigma_R\sigma_Q \qquad equation\ 6.3$$

$$\beta = \frac{\mu_m}{\sigma_m} \qquad equation\ 6.4$$

The reliability index is the number of standard deviations the mean is away from failure. For correlated R and Q, substituting from above:

$$\beta = \frac{\mu_R - \mu_Q}{\sqrt{\sigma_R^2 + \sigma_Q^2 - 2\rho_{RQ}\sigma_R\sigma_Q}} \qquad equation\ 6.5$$

For statistically independent R and Q:

$$\beta = \frac{\mu_R - \mu_Q}{\sqrt{\sigma_R^2 + \sigma_Q^2}} \qquad equation\ 6.6$$

The reliability index, β, is the distance between the mean and the failure point M=0 in units of standard deviation as shown in Figure 6.3. It is a measure of how far away the most likely value is from failure.

The area under the probability distribution f(M) where M $\leq$ 0 is the probability of failure, therefore we are interested in the CDF to define the failure $[F(m) = \int_{-\infty}^{0} f(m)dm]$. Since in this derivation Q and R are normally distributed the margin of safety function is a sum of these variables, then M is normally distributed and we can use the standard normal distribution to solve this integral.

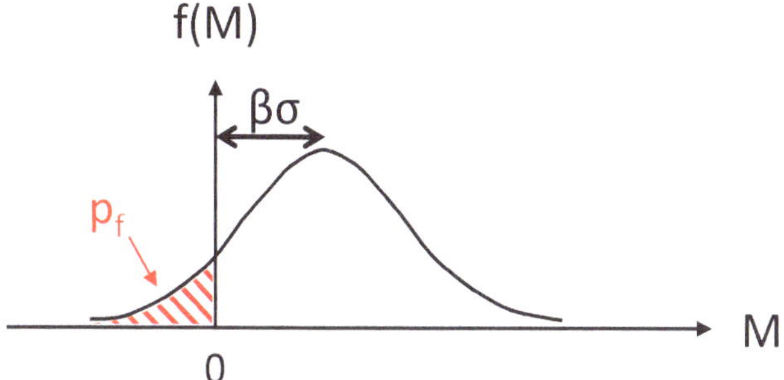

Figure 6.3 Shown is the distribution of M with the probability of failure (p_f) as the region where M ≤ 0. The distance between the failure point, M = 0, and the mean value (μ) is the reliability index (β) in units of standard deviation (σ).

$$p_f = F(M) = \Phi\left(\frac{M-\mu_m}{\sigma_m}\right) = \Phi\left(\frac{0-\mu_m}{\sigma_m}\right) = \Phi\left(\frac{-\mu_m}{\sigma_m}\right) = \Phi(-\beta) = 1 - \Phi(\beta) \qquad equation\ 6.7$$

By assuming that the load and resistance are normal and that failure is defined using the margin of safety formulation we arrive at an exact solution for the reliability index and the probability of failure. This is the most common and intuitive presentation of the component reliability calculation. Note however if load and resistance are not normal then this solution is not exact. Other solution methods exist when the assumption of normality does not hold true.

Example: Exact Solution for Normals

To demonstrate reliability using an exact solution we will look at a slope stability problem borrowed from Baecher and Christian (2003). This is called the cut slope problem because we are interested in the stability of a vertical cut in cohesive soil. Figure 6.4 shows the geometry. If we assume that the soil is purely cohesive and that the slope will fail along a 45 degree failure plane (α) then we can set up the load and resistance functions for solving the stability of this slope. For this example the height will be a deterministic value fixed at 10 m.

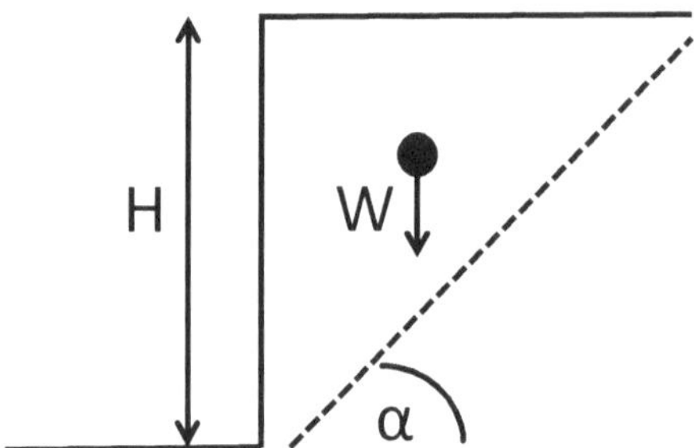

Cut slope geometry shows H is the vertical height of the cut slope, α is the angle of the failure plane, and W is the weight vector at the center of mass of the potential failure wedge.

$R = \text{c}$ cohesion of the soil

$Q = 0.25\gamma H$ from $0.5\gamma Hsin\alpha cos\alpha$ the solution for a 45° slope.

If c and γ are normally distributed then R and Q are normally distributed because both are one-to-one functions of their independent variables. The first and second moments and correlation coefficient for this problem are given as:

$\mu_c = 100\ kPa \quad \sigma_c = 30\ kPa$
$\mu_\gamma = 20\ kN/m^3 \quad \sigma_\gamma = 2\ kN/m^3$
$\rho_{c\gamma} = 0.5$

Solving for R and Q which are functions of the random variables c and γ:

$$\mu_R = 100\ kPa$$

$$\sigma_R = 30\ kPa$$

$$\mu_Q = 0.25\left(20\frac{kN}{m^3}\right)10m = 50\ kPa$$

$$\sigma_Q = \sqrt{(0.25 \cdot 10m)^2(2\frac{kN}{m^3})^2} = 5\ kPa$$

Because these are one-to-one functions relating the dependent and independent variables then R and Q have the same correlation coefficient as c and γ. Using the margin of safety formulation and having normally distributed variables lends to the exact solution:

$$M = R - Q$$

$$\mu_M = \mu_R - \mu_Q = 100 - 50 = 50\ kPa$$

$$\sigma_M^2 = \sigma_R^2 + \sigma_Q^2 - 2\rho_{RQ}\sigma_R\sigma_Q = 30^2 + 5^2 - 2(0.5)(30)(5)$$

$$= 900 + 25 - 150 = 775\ kPa$$

$$\beta = \frac{\mu_m}{\sigma_m} = \frac{50}{\sqrt{775}} = 1.80$$

(i.e., the mean is 1.8 standard deviations from failure)

$$p_f = 1 - \Phi(\beta) = 1 - \Phi(1.80) \simeq 1 - 0.96407 = 0.03593$$

So given:

- The distributions and statistics of the independent variables c and γ,
- The equations relating the independent variables to the dependent variables R and Q, and
- The margin of safety formulation defining failure,

then the probability of failure for this problem is 3.6%. The correlation between the cohesion and the density is physically obvious but let us assume that they are not correlated (i.e., statistically independent where ρ=0) to see how it influences the probability of failure.

$$\mu_M = \mu_R - \mu_Q = 100 - 50 = 50\ kPa$$

$$\sigma_M^2 = \sigma_R^2 + \sigma_Q^2 - 2\rho_{RQ}\sigma_R\sigma_Q = 30^2 + 5^2 - 0 = 900 + 25 - 0$$

$$= 925\ kPa$$

$$\beta = \frac{\mu_m}{\sigma_m} = \frac{50}{\sqrt{925}} = 1.64$$

(i.e., the mean is 1.64 standard deviations from failure)

$$p_f = 1 - \Phi(\beta) = 1 - \Phi(1.64) \simeq 1 - 0.949497 = 0.050503$$

The reliability index is lower and the probability of failure is higher for the uncorrelated situation. Why is this? In a correlated situation the uncertainties in the load and resistance are joint or interrelated, whereas the uncertainties in the uncorrelated situation are independent resulting in larger uncertainty when propagated through the margin of safety equation and therefore a smaller reliability index.

The reliability index decreases, and probability of failure increases, as the mean moves closer to the failure point. The mantra "Bigger Beta is Better" is a useful one to remember in engineering. It is common to design engineered features to have a reliability index of 2 to 3; that is a design with the mean 2 to 3 standard deviations away from the failure point. This concept will be discussed in more detail in the chapter on reliability-based codes (Chapter 9).

Lognormal Parameters

An alternate exact solution exists if load and resistance are assumed lognormal (i.e., the natural log of load and resistance is normal) and the factor of safety formulation is used to define failure.

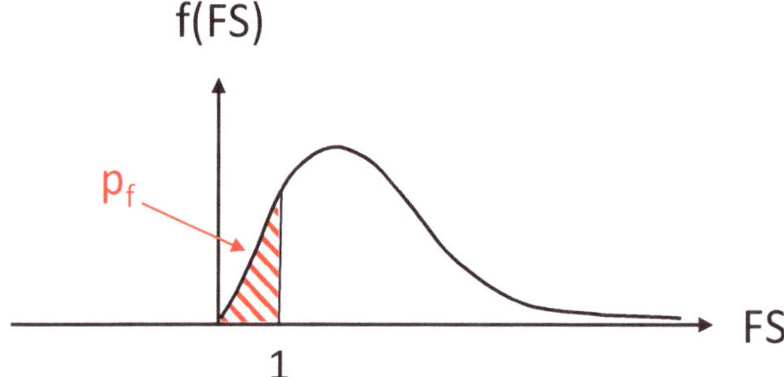

Figure 6.4 Reliability using the factor of safety formulation.

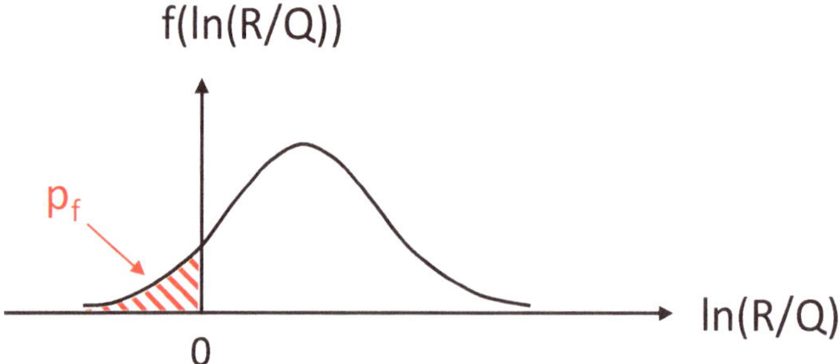

Figure 6.5 For lognormal load and resistance then the natural log of the FS=R/Q is normally distributed.

This formulation is used often for situations where the load and resistance are both non-negative, and historically was developed for steel construction.

$$FS = R/Q \quad \textit{where R and Q are jointly normal} \qquad \textit{equation } 6.8$$

By taking the natural log of both sides we find:

$$ln(FS) = ln(R) - ln(Q) \qquad \textit{equation } 6.9$$

$$\lambda_{FS} = \lambda_R - \lambda_Q \qquad \textit{equation } 6.10$$

$$\xi_{FS}^2 = \xi_R^2 + \xi_Q^2 - 2\rho_{lnRlnQ}\xi_R\xi_Q \qquad \textit{equation } 6.11$$

$$\beta = \frac{\lambda_{FS}}{\xi_{FS}} \qquad \textit{equation } 6.12$$

$$p_f = F(1) = \Phi\left(\frac{ln(1) - \lambda_{FS}}{\xi_{FS}}\right) = \Phi\left(\frac{0 - \lambda_{FS}}{\xi_{FS}}\right)$$
$$= \Phi\left(\frac{-\lambda_{FS}}{\xi_{FS}}\right) = \Phi(-\beta) = 1 - \Phi(\beta) \qquad \textit{equation } 6.13$$

The reliability index can also be written as:

$$\beta = \frac{\mu_{lnR} - \mu_{lnQ}}{\sqrt{\sigma_{lnR}^2 + \sigma_{lnQ}^2 - 2\rho_{lnRlnQ}\sigma_{lnR}\sigma_{lnQ}}}$$
$$= \frac{ln\left(\frac{\mu_R}{\mu_Q}\sqrt{\frac{1+\delta_Q^2}{1+\delta_R^2}}\right)}{\sqrt{ln(1+\delta_R^2)ln(1+\delta_Q^2) - 2ln(1+\rho_{RQ}\delta_R\delta_Q)}} \qquad \textit{equation } 6.14$$

where the mean and standard deviation of the natural log are shown below for R, and follow the same pattern for Q.

$$\mu_{lnR} = \lambda_R = ln\mu_R - \frac{1}{2}\xi_R^2 \qquad \textit{equation } 6.15$$

$$\sigma_{lnR}^2 = \xi_R^2 = ln\left(1 + \frac{\sigma_R^2}{\mu_R^2}\right) = \ln(1 + \delta_R^2) \qquad \textit{equation } 6.16$$

An approximation of the reliability index that is commonly used for the lognormal case with given mean and coefficient of variations (Rosenblueth and Esteva, 1972) can be written as:

$$\beta \simeq \frac{\ln\left(\frac{\mu_R}{\mu_Q}\right)}{\sqrt{\delta_R^2 + \delta_Q^2 - 2\rho_{RQ}\delta_R\delta_Q}} \qquad \textit{equation } 6.17$$

General Reliability Procedure

Most reliability problems require more sophisticated solutions because the distributions of load and resistance cannot be assumed to be normal (or lognormal). The various solutions however all follow similar steps;

1. Determine the equations, formulas, models, that will be used to calculate R and Q. These can be empirical, theoretical, or approximate.
2. Calculate the first and second moments of R and Q. Mean and coefficient of variation are often sufficient, but the full distributions can be used if available.
3. In most cases the margin of safety formulation is used, $M = R - Q$ so the first and second moments of M are calculated. Here the uncertainty from R and Q are propagated to M.
4. Calculate the reliability index, **β**.
5. Calculate the probability of failure, p_f.

The following is a list of common reliability solution techniques used in Civil Engineering. The asterisk (*) indicates a method that is covered in this text.

- Exact solutions* as described above. These solutions are limited by the assumption of normal or lognormal R and Q and the corresponding failure formulation.
- First Order Second Moment (FOSM)* applies to any distribution of R and Q, but is only approximate.
- Second Order Second Moment (SOSM) has increased accuracy over FOSM, but still an approximate.
- Point-Estimate is an interesting technique similar to Gaussian quadrature integration, but can be rather cumbersome in this age of fast computing (see Rosenbluth, 1975; and Baecher and Christian, 2003).

- First Order Reliability Method (FORM)* is the "standard" of reliability analysis and often considered requisite when doing probability of failure calculations.
- Second Order Reliability Method (SORM)* is particularly useful as the failure surface becomes more non-linear.
- Monte Carlo Simulations (MC)* is the "brute force" approach that provides a robust approximate, often used to confirm results found using other methods.

FOSM

First Order Second Moment (FOSM) is the same error propagation technique presented for functions of random variables in Chapter 5. Here the function of interest is the margin of safety formulation (or the factor of safety formulation) and the random variables are R and Q. This solutions works in any situation, but provides an approximate solution. As the functions of R and Q become more non-linear the FOSM results can diverge from the true results. The following example uses the same cut slope problem but avoids the assumption of normally distributed R and Q.

Example: FOSM Solution

$$M = R - Q$$

where $R = c$ and $Q = 0.25\gamma H$ and $H = 10m$ therefore

$$M = c - 2.5\gamma$$

$$\mu_M = \mu_c - 2.5\mu_\gamma = 100 - 2.5(20) = 50$$

$$\sigma_M^2 = \sigma_c^2\left(\frac{\partial M}{\partial c}\right)^2 + \sigma_\gamma^2\left(\frac{\partial M}{\partial \gamma}\right)^2 + 2\rho\sigma_c\sigma_\gamma\frac{\partial M}{\partial c}\frac{\partial M}{\partial \gamma}$$

$$\frac{dM}{dc} = 1 \quad \frac{\partial M}{\partial \gamma} = -2.5$$

$$\sigma_M^2 = 30^2(1)^2 + 2^2(-2.5)^2 + 2(0.5)30(2)1(-2.5) = 900 + 25 - 150$$
$$= 775$$

$$\beta = \frac{\mu_m}{\sigma_m} = \frac{50}{\sqrt{775}} = 1.80$$

$$p_f = 1 - \Phi(\beta) = 1 - \Phi(1.80) \simeq 1 - 0.96407 = 0.03593$$

For uncorrelated load and resistance:

$$\sigma_M^2 = 30^2(1)^2 + 2^2(-2.5)^2 + 2(0)30(2)1(-2.5) = 900 + 25 = 925$$

$$\beta = \frac{\mu_m}{\sigma_m} = \frac{50}{\sqrt{925}} = 1.64$$

$$p_f = 1 - \Phi(\beta) = 1 - \Phi(1.64) \simeq 1 - 0.949497 = 0.050503$$

Second Order Second Moment (SOSM) takes the approximation further using the second order expansion of the Taylor series. This may be more accurate for problems where the second partial derivative of the R and Q have non-zero results.

It should be noted that FOSM (and SOSM by extension) involve some assumptions that can lead to inaccurate results. This was first shown by Hasofer and Lind (1974), and more recently demonstrated using the cut slope problem by Baecher and Christian (2003). The previous example shows the cut slope problem with uncorrelated load and resistance using the margin of safety formulation which results in a FOSM-based reliability index of 1.64. The same problem using a factor of safety formulation will give a FOSM-based reliability index of 1.58, even though $M = 0$ and $FS = 1$ are mathematically identical. This variability of FOSM-based results as a function of the problem formulation lead researchers to investigate other invariant solutions to the reliability problem.

Monte Carlo Simulations

As with functions of random variables we can solve the margin of safety formulation (or factor of safety formulation) using Monte Carlo simulations. The MatLab script for the cut slope problem is shown below.

Example: Monte Carlo Approximation

```
%%%%%%%%%%%%%%%%%%%%%%%%%%%%%%%%%%%%%%%%%%%%%%%%%%%%%%%%%%%
%MC of vertical cut problem (correlated)
```

```
%uniformly distributed random numbers from 0 to 1
u1=unifrnd(0,1,10000,1);
u2=unifrnd(0,1,10000,1);

%simulations of c
c=norminv(u1).*30+100;
%simulations of gamma given c
gamma=(norminv(u2)).*(2).*(sqrt(10.50.^2))+((0.50).*(2/
30).*(c-100))+(20);
%margin of safety
M=c-(gamma.*(10/4));

%results
hist(M,100);
x_bar=mean(M)
s=std(M)
beta=x_bar/s
pf=normcdf(-beta,0,1)
%%%%%%%%%%%%%%%%%%%%%%%%%%%%%%%%%%%%%%%%%%%%%%%%%%%%%%%
```

The results are shown below for the correlated example.

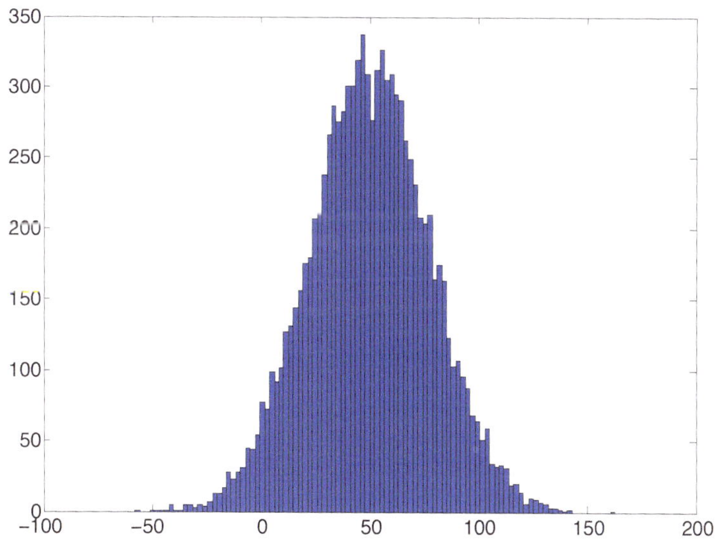


```
x_bar = 49.9724
s = 27.7358
beta = 1.8017
pf = 0.0358
```

To run the simulations for the uncorrelated example can set the correlation coefficient to zero in the previous script or rewrite the script in a much simpler manner as shown below.

```
%%%%%%%%%%%%%%%%%%%%%%%%%%%%%%%%%%%%%%%%%%%%%%%%%%%%%%%%
% MC of vertical cut problem (uncorrelated normals)

c=normrnd(100,30,10000,1);
gamma=normrnd(20,2,10000,1);

M=c-(gamma.*(10/4));

hist(M,100);
x_bar=mean(M)
s=std(M)
beta=x_bar/s
pf=normcdf(-beta,0,1)
%%%%%%%%%%%%%%%%%%%%%%%%%%%%%%%%%%%%%%%%%%%%%%%%%%%%%%%%
```

And the results for the uncorrelated load and resistance are:

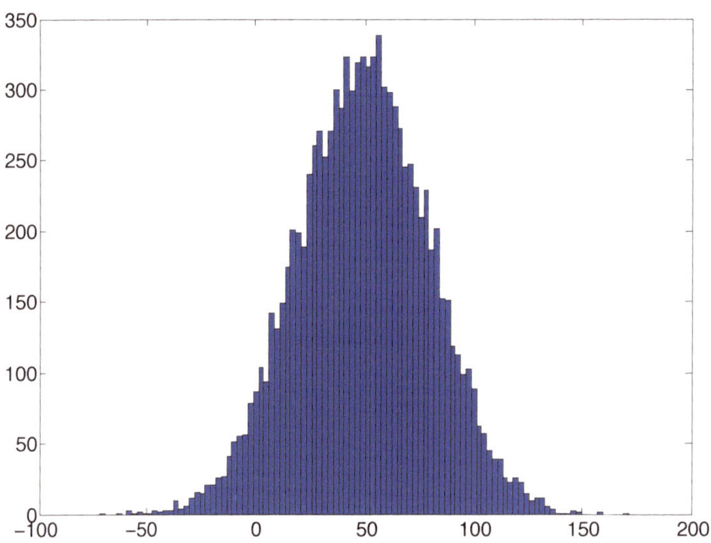


```
x_bar = 49.4460
s = 30.1155
beta = 1.6419
pf = 0.0503
```

Note that the results vary depending on how many simulations we run. In this example 10,000 simulations were run. As discussed previously we can run the number of simulations high enough to minimize the error below a certain acceptable tolerance. With Monte Carlo simulations we are estimating the results; however given enough simulations the results become fairly accurate.

FORM/SORM

A more robust means of determining the probability of failure is using the geometric approach of First Order Reliability Method (FORM) and/or Second Order Reliability Method (SORM). As discussed FOSM can give different results dependent on if the problem is formulated using margin of safety or factor of safety. To avoid this the Hasofer-Lind approach (Hasofer & Lind, 1974) to FORM translates the problem into standard normal space and then solves for the distance from the mean to the point of failure, which is invariant regardless of how failure is formulated.

The Hasofer-Lind approach typically follows:

1. Formulate the problem as margin of safety, $M = R - Q$.
2. Transform from (R,Q) space into standard normal space (R',Q') as shown conceptually in Figure 6.6. [This transformation is a function of the joint distribution of R and Q.]
3. Find minimum distance from M=0 to origin using a straight line fit at the tangent point (see Figure 6.7). This requires an iterative solution which can be accomplished using different approaches.
4. The reliability index is equal to the minimum distance to the tangent, $\beta = \min(distance)$, and the probability of failure is the standard normal distribution of the negative reliability index, $p_f = \Phi(-\beta)$.

SORM can provide a more accurate solution as it fits a curve instead of a straight line to M=0 when solving the minimum distance to the tangent (Der Kiureghian et al., 1987). Both FORM and SORM require an iterative solution to determine the minimum distance to the failure surface and the most efficient solutions use matrix manipulations (e.g., Cholesky decomposition, Jacobian matrix, etc). In Appendix C the "improved" HLRF algorithm (Zhang and Der Kiureghian, 1995) is presented for solving the cut slope problem using FORM.

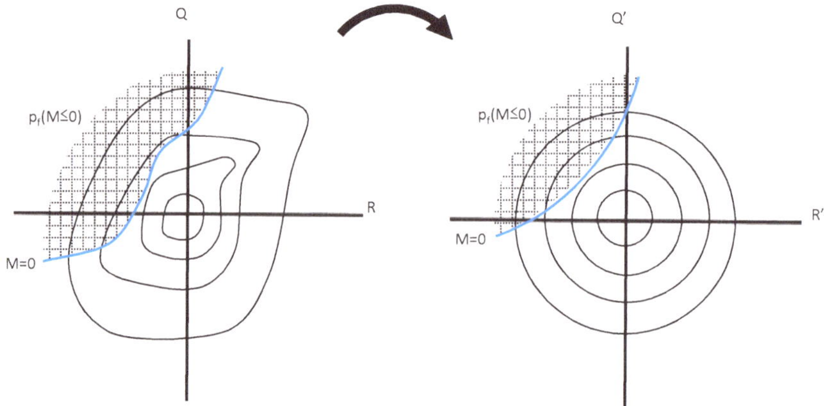

Figure 6.6 Transforming to standard normal space, from (R,Q) to (R',Q').

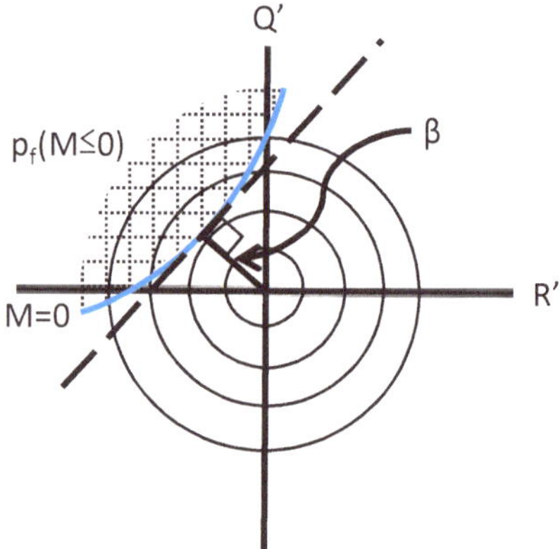

Figure 6.7 The reliability index, beta, is the minimum perpendicular distance from the origin to a straight line tangent at M=0.

To simplify the application of FORM/SORM and encourage practicing engineers to use these methods, they have been written into Excel to take advantage of the built-in matrix manipulation and iteration solver functions available in the spreadsheet program. Low and Tang (1997) programmed SORM into Excel utilizing the properties of an ellipse and built in matrix manipulation functions. These spreadsheet reliability methods provide invariant reliability solutions that are readily applicable to Civil Engineering problems.

Figures 6.8 and 6.9 show the cut slope problem solved using the Low and Tang (1997) spreadsheet solution. Notes at the bottom of the spreadsheet describe the solution method. Figure 6.10 shows the equations in each cell and how the Solver is used to iteratively calculate the minimum reliability index. The ease of this solution method renders reliability accessible to any practicing engineer with the interest in performing reliability.

Other spreadsheet solutions exist and subsequent work has expanded these solutions to accommodate many probability distributions and the transformation of those distributions into standard normal space. For further reading the following references are recommended; Low and Tang (2004), Phoon and Nadim (2004), and Low (2005).

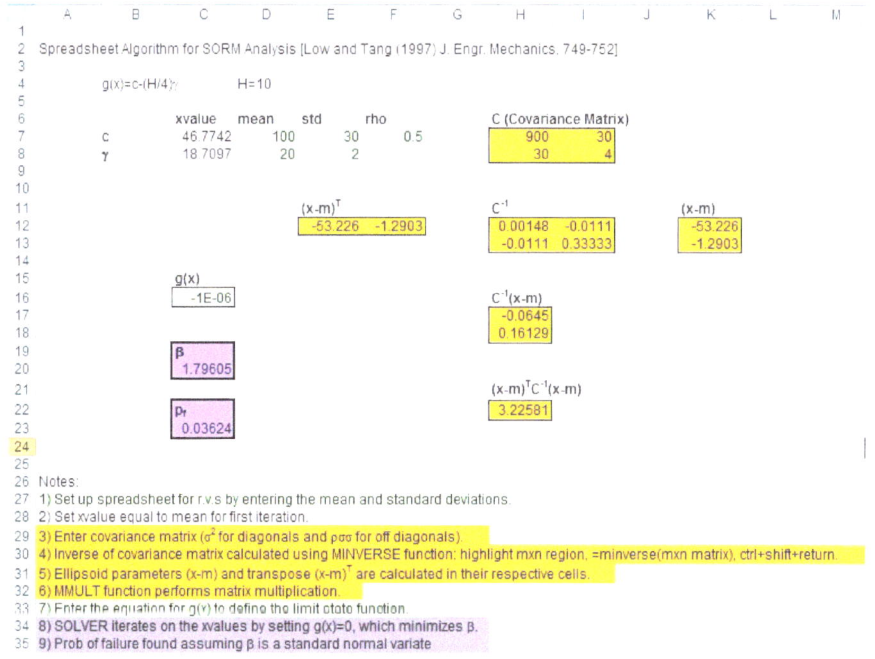

Figure 6.8. Cut slope problem with correlated load and resistance.

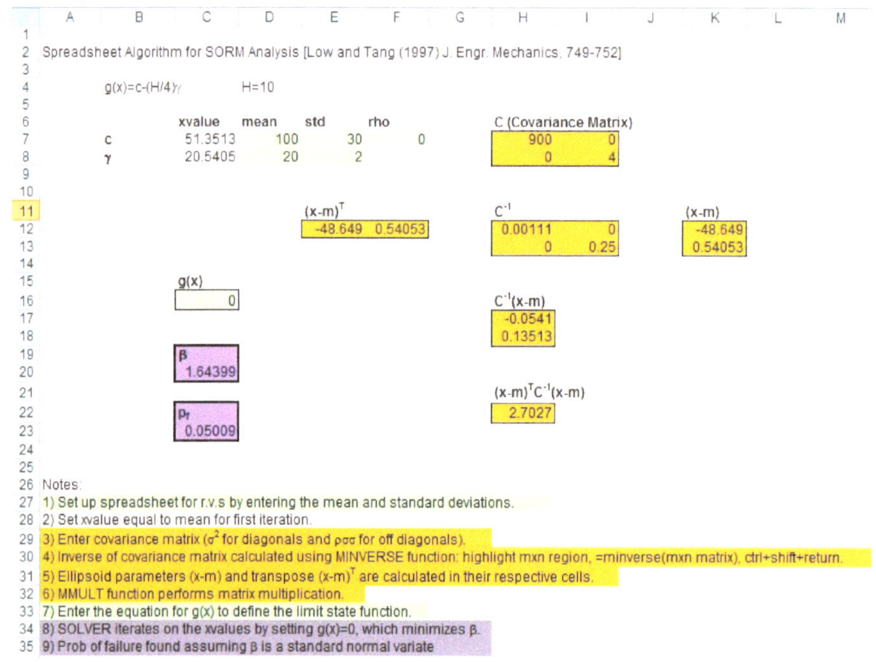

Figure 6.9. Cut slope problem with uncorrelated load and resistance.

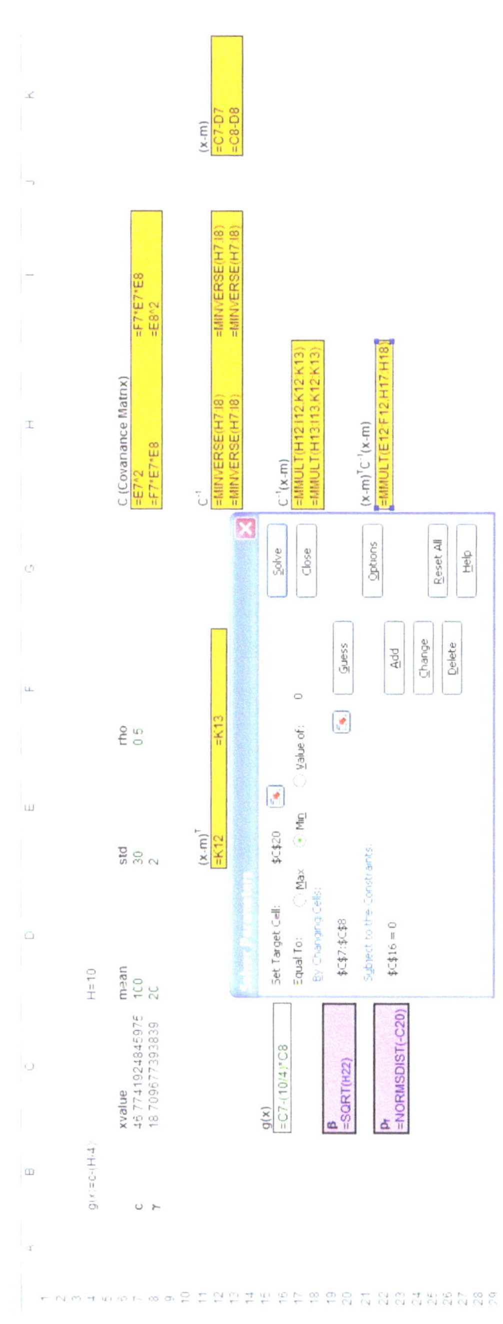

Figure 6.10 Equations and Solver for iteratively calculating the minimum reliability index using SORM spreadsheet solution.

Limit State

Up to this point we have been discussing failure and defining failure primarily using the margin of safety formulation. However we may be interested not in outright failure but some unsatisfactory performance that we wouldn't necessarily label as failure in the breaking, fracturing, or collapsing sense. In defining a performance criterion in this manner it is then often called a limit state. The same mathematics and solution techniques can be used to solve any limit state. We generalize the margin of safety formulation to now encompass a threshold beyond which we can have unsatisfactory performance. The limit state function is usually denoted by g and the independent variables X where $g \leq 0$ is unsatisfactory performance:

$$g = X_1 - X_2 \qquad equation\ 6.18$$

Chapter Summary

- Reliability is solving a specific function of random variables where the variables are **load** and **resistance** and the function is commonly the difference of the two (i.e., margin of safety formulation).
- Reliability is described by the **reliability index**, the number of standard deviations away from failure.
- The reliability index can be related mathematically to the **probability of failure**.
- The solution techniques covered in the previous chapter (Chapter 5 Functions of Random Variables) can be applied in solving reliability problems; **exact solutions** for normal and lognormal, **approximate solutions** using a Taylor series expansion, and **Monte Carlo** simulations.
- Additionally, geometric solution techniques specific to reliability, **FORM** and **SORM**, are used because they are invariant with respect to the function that relates load and resistance.
- A simplified SORM spreadsheet solution is presented to make reliability accessible without the need for a background in matrix mathematics.
- Reliability is generalized to the **limit state** formulation for solving problems involving any unsatisfactory engineering performance, not just failure.
- In engineering the goal is to achieve a safe design that is also economical. The reliability index (β) is the measure of how many standard deviations away from failure the design is, therefore within the economic constraints of the project, **Bigger Beta is Better**.

7 SYSTEM RELIABILITY ANALYSIS

Component reliability was covered in Chapter 6, but we may often have a system made of many components. To analyze a system we can generalize and expand the limit state by considering multiple components and multiple failures that define the system [we will be using the term failure as synonymous with any unacceptable performance as defined by the limit state].

$g = X_1 - X_2$ where $g \leq 0$ is failure or unsatisfactory performance

$g_j(\boldsymbol{X}) = g(x_1, x_2, \ldots, x_n)$ where $j = 1, 2, \ldots, k$ so there are k potential failures to evaluate.

If we restrict the system to just two independent variables and three limit states [$\boldsymbol{X} = (x_1, x_2)$ and $j = 1,2,3$ where $g_j(\boldsymbol{X}) = 0$] we can visualize the multiple limit states in standard normal space as shown in Figure 7.1.

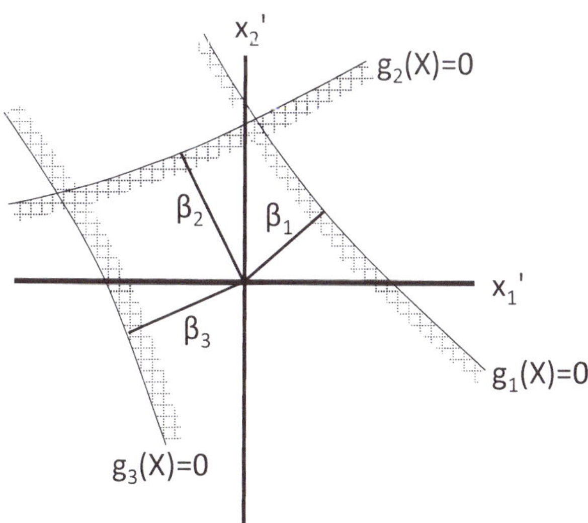

Figure 7.1 Multiple limit states, $g_j(X)$, defining system failure, $g_j(X)=0$.

If the joint PDF of the independent variables is $f(x_1, \ldots, x_n)$ and any component failure results in system failure then the probability of failure for the system in general is the volume integral:

$$p_f = \int_{g(\boldsymbol{X})\leq 0} \cdots \int f(x_1, \ldots, x_n) dx_1 \ldots dx_n \qquad equation\ 7.1$$

Calculation of this multifold integral can often be difficult and in most cases approximations or bounds are used to estimate the range of the probability of system failure.

Idealized Systems

We often idealize a system into series or parallel to aid in conceptualizing how components are connected and interrelated.

Series

A series system is one where the entire system fails if any component in the systems fails (i.e., "weak-link-in-the-chain"). This is a non-redundant system and is the type of system that engineers should avoid; unfortunately in Civil Engineering we build many systems that are in series, particularly lifelines such as; highways, bridges, pipelines, communication/transmission lines, canals, levees, etc. And even when systems are redundant at one scale, they can often be non-redundant at a different scale (e.g., the electricity grid has redundancy at the local level but when scaled to the regional level is often dependent on a single component or node when connecting to a neighboring regional grid).

The probability of failure of a series system is the joint union of all the failure states of the components.

$$p_{f,series} = P\left[\bigcup_{j=1}^{k} \left(g_j(X) \leq 0\right)\right] \qquad equation\ 7.2$$

Parallel

Parallel systems are redundant systems that only fail when all components in the system fail. These are desirable qualities for an engineered system. Redundancy is often critical when we are addressing collapse or other catastrophic failure modes. The probability of failure of a parallel system is the joint intersection of all the failure states.

$$p_{f,parallel} = P\left[\bigcap_{j=1}^{k} \left(g_j(X) \leq 0\right)\right] \qquad \textit{equation } 7.3$$

General

Some systems can be a combination of series and parallel components in different arrangements (e.g., a water distribution system is better characterized as a general system). To generalize the discussion we can talk about cut sets and link sets; a cut set results in system failure and a link set results in system survival.

- In a series system each component is a cut set and all components together is a link set.
- For a parallel system, all components together form the only cut set and every component is a link set.

The probability of failure for a general system can be described as follows where the cut sets are defined as $[C_1 \ldots C_M]$.

$$p_{f,general} = P\left[\bigcup_{m=1}^{M} \bigcap_{j \in C_M} \left(g_j(X) \leq 0\right)\right] \qquad \textit{equation } 7.4$$

Here we are taking the intersection of the events within the cut sets because these are parallel, and the union of the events between cut sets because these are series.

Example: Two Component System

The following example evaluates a simple two component system arranged in series and parallel. The two components have the same component probability of failure. The component failures are correlated due to similar construction and materials and/or similar loading:

$$P(A) = P(B) = 0.1 \qquad P(A|B) = 0.5$$

If we arrange the components into a series system, then the probability of failure would be the union of the failure events; the failure of component A **or** the failure of component B:

$$
\begin{aligned}
P(A \cup B) &= P(A) + P(B) - P(AB) && \textit{addition rule} \\
&= P(A) + P(B) - P(A|B)P(B) && \textit{multiplication rule} \\
&= 0.1 + 0.1 - (0.5)0.1 = 0.15
\end{aligned}
$$

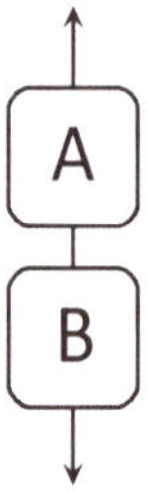

Notice that the correlation between the two components reduces the probability of failure of a series system. If the component failures were perfectly correlated the system probability of failure would be 0.1, and if they were statistically independent the system probability of failure would be 0.19 (calculate these yourself to verify).

If we arrange the two components in a parallel system then the system probability of failure would be the intersection of the failure events; the failure of component A **and** the failure of component B:

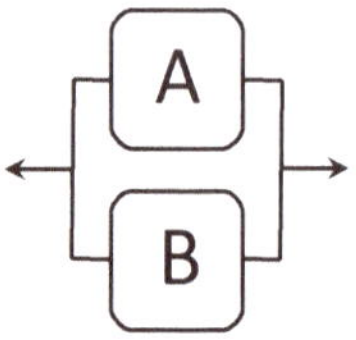

$$
\begin{aligned}
P(AB) &= P(A|B)P(B) && \textit{multiplication rule} \\
&= (0.5)0.1 = 0.05
\end{aligned}
$$

If the components were perfectly correlated the system probability of failure would be 0.1, and if they were statistically independent the system probability of failure would be 0.01 (calculate these yourself to verify). Notice that correlation increases the probability of failure of parallel systems.

System Bounds

Bounds on the system probability of failure provide a simplified way of estimating the range that the probability of failure can take. How narrow the estimate of the range is a function of the complexity of the bounds estimate and how much information is included in estimating the bounds. The first order or unimodal bounds neglects the specific correlation between events and therefore can be rather wide. Higher order bounds account for specific correlation between events and provide narrower estimates as a function of the degree of correlation included.

Unimodal Bounds for Series Systems

Unimodal Bounds for positively correlated $(\rho_{ij} > 0)$ individual failure events $(E_i = [g_i(x) < 0])$ in a series system can be written as:

$$(\max_i P(E_i)) \leq \; p_{f,series} \; \leq \left(1 - \prod_{i=1}^{k} (1 - P(E_i))\right)$$

$$\simeq \sum_{i=1}^{k} P(E_i) \; for\; small\; component\; p_f$$

$$equation\; 7.5$$

The left side of the inequality states that the lower unimodal bound on the probability of failure for a series system with positively correlated failure events is the max of any single failure event. Whereas right side of the inequality states that the upper bound is approximately the sum of the probability of failure of all failure events, as the positive correlation between events will result in something less than this.

The complimentary probability of safety ($p_s = 1 - p_f$) is then:

$$\prod_{i=1}^{k} P(\bar{E}_i) \leq \; p_{s,series} \; \leq \min_i P(\bar{E}_i) \qquad equation\; 7.6$$

The left side of this inequality is the product of the compliment of the probability of failure of all failure events, and the right side is the minimum of the compliment of the probability of failure for any single failure event.

Unimodal bounds for negatively correlated failure events $(\rho_{ij} < 0)$ in a series system are:

$$\left(1 - \prod_{i=1}^{k} P(\bar{E}_i)\right) \leq \; p_{f,series} \; \leq 1 \qquad equation\; 7.7$$

The left side of the inequality is the compliment of the product of all non-failures, and the right side is one (which is rather uninformative). The probability of safety is then:

$$0 \leq \; p_{s,series} \; \leq \left(\prod_{i=1}^{k} P(\bar{E}_i)\right) \qquad equation\; 7.8$$

The left side of the inequality is zero (again rather uninformative), and the right side is the product of the compliment of the probability of failure for all failure events.

Example: Reservoir Problem

A reservoir is designed for both flood control and water supply. Flood control (F) is affected by snow melt (A) and rainfall (B). And a low reservoir (G) can be caused by a dry winter (C) and low rainfall (D). Say we know that snow melt and rainfall are positively correlated (i.e., wet winters lead to wet springs) but don't have a value of the correlation coefficient (ρ_{AB}). And we know that dry winters and low rainfall are positively correlated (i.e., dry winters lead to dry springs) but again we don't have a value of the correlation coefficient (ρ_{CD}). Intuitively we know that flooding and drought are negatively correlated (ρ_{FG}). If we are given the following component probabilities what is the probability of poor reservoir performance?

$$P(A) = 0.15 \;\; P(B) = 0.20 \;\; P(C) = 0.10 \;\; P(D) = 0.20$$

$$p_f = P(F \cup G) = P((A \cap B) \cup (C \cap D))$$

The statement above reads; the probability of failure or probability of poor performance of the reservoir can be due to flood control problems **or** a low reservoir. Flood control problems can be caused by snow melt **and** rainfall, and a low reservoir can be caused by low rainfall **and** a dry winter. We can estimate the bounds of this system using a first-order or unimodal approximation.

$p_{f,series} \geq 1 - P(\bar{F})P(\bar{G})$ because F and G are negatively correlated

To determine $P(\bar{F})$ we use unimodal bounds for positively correlated events.

$$\begin{aligned} [P(\bar{A})P(\bar{B})] \leq \;& P(\bar{F}) \leq [\min(P(\bar{A}), P(\bar{B}))] \\ 0.85 \cdot 0.80 \leq \;& P(\bar{F}) \leq 0.80 \\ 0.68 \leq \;& P(\bar{F}) \leq 0.80 \end{aligned}$$

Similarly solving for $P(\bar{G})$.

$$\begin{aligned} [P(\bar{C})P(\bar{D})] \leq \;& P(\bar{G}) \leq [\min(P(\bar{C}), P(\bar{D}))] \\ 0.90 \cdot 0.80 \leq \;& P(\bar{G}) \leq 0.80 \\ 0.72 \leq \;& P(\bar{G}) \leq 0.80 \end{aligned}$$

Therefore

$$p_{f,series} \geq 1 - 0.80 \cdot 0.80 = 0.36$$

The probability of unsatisfactory reservoir performance is greater than or equal to 36% which is information that can aid in the decision or planning process for the reservoir as part of a larger a water system.

Bimodal Bounds for Series Systems

To improve the bounds (i.e., find a narrower approximation of the bounds) on a series system we can use bimodal or higher estimates. Bimodal bounds partially account for the correlation between failure events by using event pairs or joint events (E_iE_j). Ordering of the events can affect the results with higher modal estimates, and those interested are encouraged to read Song and Der Kiureghian (2003) for further details.

A method of bimodal bounds for series systems was derived by Kounias (1968) and Hunter (1976) and simplified by Ditlevsen (1979) by assuming Gaussian variates:

$$P(E_1) + \sum_{i=2}^{k} max\left(P(E_i) - \sum_{j=1}^{i-1} P(E_iE_j)\,; 0 \right) \leq \ p_{f,series}$$

$$\leq P(E_1) + \sum_{i=2}^{k}\left(P(E_i) - \max_{j<i} P(E_iE_j)\right) \quad equation\ 7.9$$

Example: Unimodal vs. Bimodal

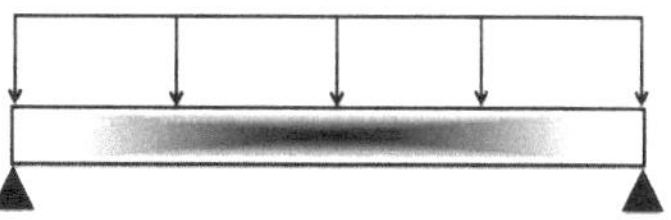

A simply supported beam subjected to a uniformly distributed load can fail in flexure (f_1), shear (f_2), or both flexure and shear (f_3). Therefore failure of this system happens if any component fails $p_{f,series} = P(f_1 \cup f_2 \cup f_3)$. Component reliability has been performed for each failure mode and the results are;

	β_i	$p_{s,i}$	$p_{f,i}$
f_1	1.59	0.9445	0.0555
f_2	1.57	0.9418	0.0582
f_3	1.57	0.9418	0.0582

The failure modes are assumed to be positively correlated. The unimodal bounds of this system are:

$$\max(0.0555{,}0.0582) \leq p_{f,series} \leq 1 - (0.9445 \cdot 0.9418 \cdot 0.9418)$$
$$\simeq (0.0555 + 0.0582 + 0.0582)$$

$$0.0582 \leq p_{f,series} \leq 0.1622$$
$$\simeq 0.1719$$

If some numerical simulations (e.g., finite element) were run to determine how the failure modes are correlated we may have the following results:

$$P(f_1|f_2) = 0.45; \;\; P(f_1|f_3) = 0.99; \;\; P(f_2|f_3) = 0.57$$

The joint probabilities are then:

$$P(AB) = P(A|B)P(B) \quad multiplication\, rule$$
$$P(f_1 f_2) = 0.0250; \quad P(f_1 f_3) = 0.0576; \quad P(f_1 f_2) = 0.0332$$

The bimodal lower bound is then:

$$P(f_1) + \big(max\big[\big(P(f_2) - P(f_1 f_2)\big); 0\big]$$
$$+ \max\, [(P(f_3) - P(f_1 f_3) - P(f_2 f_3)]); 0\big)$$
$$0.0555 + (max[(0.0582 - 0.0250); 0]$$
$$+ \max\, [(0.0582 - 0.0576 - 0.0332]); 0)$$
$$0.0887$$

And the upper bound is:

$$P(f_1) + [P(f_1) - P(f_2 f_1)] + [P(f_3) - P(f_3 f_1)]$$
$$0.0555 + [0.0582 - 0.025] + [0.0582 - 0.0576]$$
$$0.0893$$

Therefore:

$$00887 \leq p_{f,series} \leq 0.0893$$

The table below shows the comparison between unimodal and bimodal bounds.

	Lower Bound	**Upper Bound**
Unimodal	0.0582	0.1622
Bimodal	0.0887	0.0893

In this example the bimodal bounds provided a much narrower range by including the correlation of joint events. The bimodal bounds here may constrain the system probability of failure sufficiently that it can be useful for engineering decision purposes.

Unimodal Bounds for Parallel Systems

Estimating the probability of failure for a parallel systems is often not as critical as a series system because parallel systems have redundancy. Nonetheless the first order bounds for a parallel system treats the components as the wholly uncorrelated or perfectly correlated:

$$\prod_{i=1}^{k} P(E_i) \leq \; p_{f,parallel} \; \leq \min_i P(E_i) \qquad equation\ 7.10$$

These bounds are often very wide and not highly informative. For parallel systems with a small number of components an estimate can be accomplished by using higher order bounds or multifold integration (Ang and Tang, 1984).

For structures or other built features redundancy of the components is most often **active** as the components are each carrying load even though they are in a parallel arrangement. An example of active redundancy are structural columns that are designed to carry the full load in the event of failure of a nearby column. This compares to redundancy that is passive where parallel components are standby or backup in case of emergency. An example of **passive** redundancy is backup generators at a hospital that come online when there is a power outage. Parallel systems with active redundancy are markedly different than those with passive redundancy and should be treated accordingly.

Correlation and Components

Correlation between components can have a dramatic impact on the probability of failure, the difficulty in evaluating this impact often arises in the measuring and accounting for the correlation. Also the number of components in a system can have a large impact on the system probability of failure. The following tables show the general trends that correlation and number of components have on the probability of failure for a series or parallel system (Hollenback, 2013). To more formally calculate this impact some type of simulation is often required. Alternatively, higher order bounds can provide an estimate.

If a series system has positive correlation between components then it is more reliable than a system of statistically independent components. Positive correlation is beneficial to series systems because if one component is in survival then all of the components are likely to be in survival, which leads to system survival. Since the same holds true for one component being in failure state, it might seem that the beneficial effect of positive correlation would be cancelled out. However, if one component is in failure then the system fails regardless of other component states.

Table 7.1: Table showing the general effect that correlation has on series and parallel systems.

	Inc +ρ	**Dec +ρ**
Series System	↓ Dec p_f	↑ Inc p_f
Parallel System	↑ Inc p_f	↓ Dec p_f

If a parallel system has positive correlation between components it will be less reliable than a system of statistically independent components. Positive correlation is detrimental to parallel systems because if one component is in failure then all of the components are likely to fail, which leads to system failure. Since the same holds true for one component being in survival state, it might seem that the detrimental effect of positive correlation would get cancelled out. However, if only one component is in survival state then the system survives regardless of the other component states.

Table 7.2: Table showing the general effect that the number of components has on series and parallel systems.

	Inc # Components	**Dec # Components**
Series System	↑Inc p_f	↓Dec p_f
Parallel System	↓Dec p_f	↑Inc p_f

When considering the impact of the number of components of an idealized system we must consider system redundancy. Series systems are non-redundant, therefore any increase in the number of components increases the likelihood of system failure. Parallel systems however are redundant and increasing the number of components increases the redundancy thereby decreasing the probability of failure.

Defining components of a system can be straight forward for some systems, and ambiguous for other systems. For a bridge system composed of deck sections the components are obvious. The same for a structure with columns supporting a floor slab. For systems such as highways or levees, determining what constitutes a component can be difficult and somewhat arbitrary. For spatially distributed systems such as lifelines we can define a component based on its spatially correlated load and resistance (Hollenback and Moss, 2011; Moss and Germeraad, 2013).

Example: Parallel and Series Power Generators

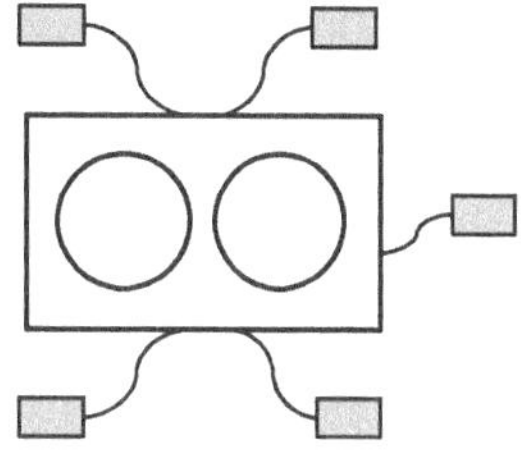

A nuclear power plant has five redundant backup power generators. These are designed to withstand strong ground shaking. Any one of the generators can provide sufficient energy to safely shut down the power plant, so if they are wired independently and are distributed around the site then this system might be considered a parallel system.

The component load and resistance, in typical ground shaking units of gravity, are assumed normally distributed with the following moments:

$$Q = N(0.10g, 0.03g) \quad R_i = N(0.20g, 0.05g)$$

The load is the forecast of strong ground shaking associated with some future event (usually estimated in a probabilistic manner). The resistance is the estimated capacity a generator has to resist the strong ground shaking. The limit state of each component is then:

$$g_i(X) = R_i - Q$$

If we assume that load and resistance are statistically independent the component probability of failure is:

$$p_{f_i} = \Phi\left(-\frac{0.20g - 0.10g}{\sqrt{0.05g^2 + 0.03^2}}\right) = 1 - \Phi(1.71) \cong 0.044$$

The system probability of failure for this parallel system is the intersection of all failure events (E_i) because for the system to fail all components must fail:

$$p_f = P(E_1 \cap E_2 \cap \ldots \cap E_5)$$

The bounds on this parallel system are (Equation 7.10):

$$\prod_{i=1}^{k} P(E_i) \leq \; p_{f,parallel} \; \leq \min_i P(E_i)$$

$$1.65 \cdot 10^{-7} \leq p_f \leq 0.044$$

The left side of the inequality is for statistically independent component failure events, and the right side is for perfectly correlated component failure events. These bounds are rather large, so Ang and Tang (1984) considered the same problem and included correlation of the ground motion. This is reasonable since strong ground motions can be spatially correlated over typical distances of a power plant footprint. Ang and Tang solved for the correlation coefficient between component limit states ($\rho_{ij} = 0.265$) and then used a multifold integration approach to evaluate the system probability of failure. The exact solution is a five fold integral of the joint standard normal PDF:

$$p_f = \int_{-\infty}^{-\beta} \int_{-\infty}^{-\beta} \dots \int_{-\infty}^{-\beta} f_{g_1', g_2', \dots g_5'} \, dg_1' \dots dg_5'$$

where to get into standard normal space the limits states are written as:

$$g_i' = \frac{g_i - \mu_{g_i}}{\sigma_{g_i}}$$

To calculate this multifold integral a numerical solution is often the most tractable approach. Ang and Tang used numerical quadrature and found the system probability of failure to be $p_f = 1 \cdot 10^{-4}$. If correlation between the resistance of the individual components (due to manufacturing similarities or other) is subsequently included, then the system probability of failure would show additional increase.

Now if we evaluate the same five backup power generators for tsunami loading there are other considerations. The component load and resistance, in units of wave height, are again assumed normally distributed:

$$Q = N(0.5m, 0.5m) \quad R_i = N(1.0m, 0.2m)$$

If we assume that load and resistance are statistically independent the component probability of failure is:

$$p_{f_i} = \Phi\left(-\frac{1.0m - 0.5m}{\sqrt{0.2m^2 + 0.5m^2}}\right) = 1 - \Phi(0.928) \cong 0.1762$$

For the loading we might assume that the wave height is nearly perfectly correlated across the site ($\rho_{Q_iQ_j} \approx 1.0$) because of the scale and duration of the wave with respect to the scale of the power plant. Instead of resorting to a complex multifold integral we could use a proxy solution by assuming that the system is now acting in series, if one generator fails due to a tsunami wave then all will fail because of the load correlation. Utilizing bounds of a series system (Equation 7.5):

$$(\max_i P(E_i)) \leq \; p_{f,series} \; \leq \left(1 - \prod_{i=1}^{k} (1 - P(E_i))\right)$$

$$0.1762 \; \leq \; p_f \; \leq \; 0.6206$$

The right side of the inequality is the perfectly correlated bound which is a reasonable estimate of how the system will behave given the loading of a tsunami wave.

Chapter Summary

- A system is a group of components that make up some engineered feature.
- The probability of failure of a system can be difficult to determine, often requiring multifold integration or simulations.
- Systems can be idealized as **series** systems, where the system fails when any single component fails, or **parallel** systems, where the system fails only if every component fails.
- Bounds can provide an estimate on the upper and lower range of system probability of failure. Presented are **unimodal** series and parallel system bounds and **bimodal** series system bounds.
- **Correlation** of components can have a strong influence on the system probability of failure. Increasing positive correlation between components has a beneficial impact on series systems by increasing the likelihood of overall system survival, and detrimental impact on parallel systems by increasing the likelihood of overall system failure.
- An increasing **number of components** in a system increases the probability of failure for a series system because there are more components to fail, yet decreases the probability of failure for a parallel system because of redundancy.
- Redundancy in parallel systems can be **active** or **passive**. Passive redundancy is usually in the form of a backup system that comes into play when failure occurs. Active redundancy is often in the form of load bearing components that are designed to carry additional load in event of nearby component failure.

8 INTRODUCTION TO DECISION ANALYSIS

Component (Chapter 6) and System (Chapter 7) reliability have been presented for computing the probability of failure, but engineering problems are often more complicated than calculating the component or system reliability. In this chapter some tools for intuitively mapping out a complex engineering problem and moving towards a decision are demonstrated. These tools can be described as trees because visually they are often presented like branches of a tree. These tools are grouped here into four types; Event Trees, Decision Trees, Logic Trees, and Fault Trees. Schematics of these trees are shown in Figures 8.1 and 8.2.

An Event Tree maps the potential outcomes of some situation with the probability of those outcomes explicitly described on each branch. The probabilities must sum to 1.0 vertically down across all branches thereby capturing the total probability at each step. Decision Trees are Event Trees with the costs or consequences for each branch included. The probability of a particular outcome is multiplied by the cost of that outcome thereby quantifying the risk of that particular outcome. Decision Trees provide risk, a calibrated metric, for engineering decisions as was first discussed in Chapter 1. The probability of each branch can be determined using the various tools described throughout this text; statistics, probability, and reliability. Often one branch of an Event or Decision Tree requires a component (or system) reliability analysis to define the probability of that particular outcome.

A Logic Tree is often used in situations where a particular problem has multiple solution methods. Rather than arbitrarily selecting one solution method, all methods are used and weighted according to the accuracy of or confidence in a particular method. The probabilities here are often determined using expert consensus. Lacking any prior information equal probabilities should be used. Sensitivity studies can then determine the impact of a particular method on the results. A Logic Tree provides a robust mean or median estimate by using all available information to defeat epistemic uncertainty and solve the problem. However, the variance from a Logic Tree is ill defined and there is little agreement as to how best to measure the variance within this type of weighting scheme.

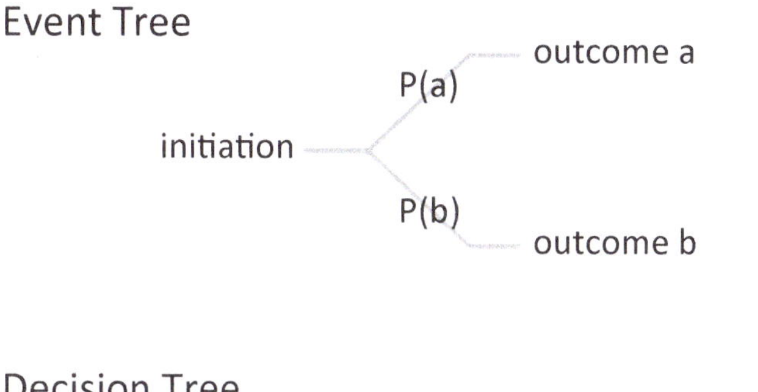

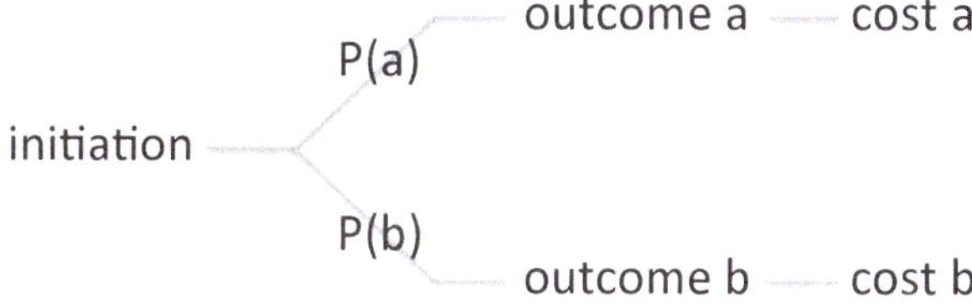

Figure 8.1 Schematics of Event and Decision Trees.

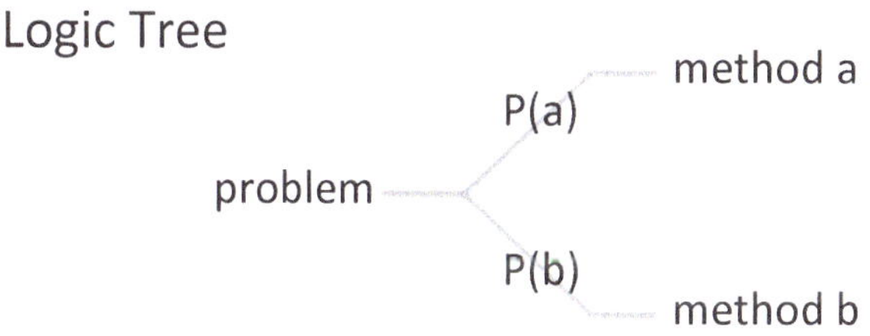

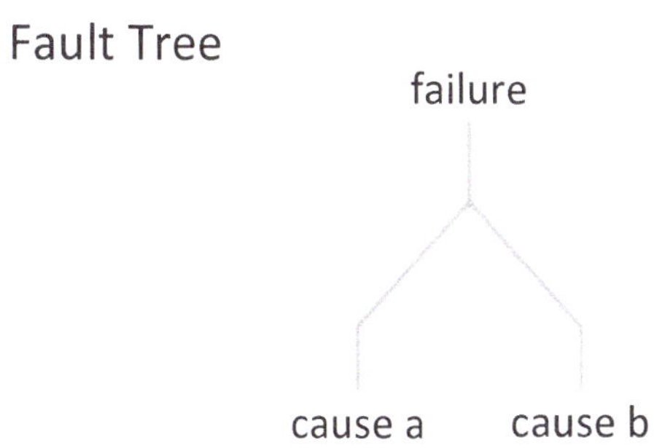

Figure 8.2 Schematics of Logic and Fault Trees.

A Fault Tree does not contain probabilities for each branch because it is a visual tool used for building often non-trivial probability statements by combining the unions and intersections of subevents that cause failure. This tree is useful for complex interdependent systems where a total probability statement is not intuitive from the outset.

The best way to fully demonstrate these tools is through example, so the bulk of this chapter will present specific problems where trees are an asset in problem solving and moving the decision process forward.

Correlation in Decision Analysis

Correlation as a concept is rather straightforward, but dealing with correlation in a specific and quantifiable manner can be at times quite difficult. The term correlation can describe many different interrelationships and phenomena. The probability of each branch of a tree can be correlated via; causal dependence, probabilistic correlation, spatial/temporal autocorrelation, and/or statistical correlation. This list may not be complete but tries to encompass the bulk of correlation. Below is described each of these types of correlation with examples from a levee system to provide concrete context.

Causal dependence is where one event causes another, this is sequential failure which is by nature a conditional probability (e.g., foundation settlement of a levee increases the probability of a flood wave overtopping the levee).

Probabilistic correlation is where two uncertainties may depend on a third uncertainty (e.g., the density of soil is an independent variable that can impact two other variables important to levee performance, soil piping and excess pore pressure generation).

Spatial or temporal autocorrelation is where two uncertainties are a function of space or time (e.g., various soil properties of the levee foundation material are spatially correlated due to the depositional nature of the soil. Seismic loading of a levee system is temporally correlated due to the finite nature of earthquake fault rupture).

Statistical correlation is where two uncertainties are estimated from a data set that is influenced by a common variance (e.g., the cohesion and friction angle of levee embankment and levee foundation soil are estimated through a linear regression of the Mohr-Coulomb failure envelope to lab data, and are negatively correlated).

Decision Analysis Examples

The following are examples demonstrating the utility of different decision tools.

Example: Power Failure Decision Tree

A power company has the option to install a backup power system at a cost of $2k per year. If power goes out the company will take a $10k hit due to penalties and fines. Based on an internal reliability analysis of the system the annual probability of failure is 10%.

install backup event E_1
- 10% failure — $2k
- 90% no failure — $2k

no backup event E_2
- 10% failure — $10k
- 90% no failure — $0k

Calculating the risk of events E_1 and E_2:

$$R(E_1) = 0.1(\$2k) + 0.9(\$2k) = \$2k$$
$$R(E_2) = 0.1(\$10k) + 0.9(\$0k) = \$1k$$

Based on this risk assessment the least expensive option would be to not install the backup system. Of course this assumes that the costs and the probabilities have been accurately assessed. [As a side note, previous case histories have demonstrated time and again that people are notoriously bad at estimating the consequences of a failure or catastrophe before hand (Moss & Germeraad, 2013), therefore in this problem the $10k hit the company may experience if no backup is installed may be an optimistically low estimate of the consequences.]

Example: Shell Mound Decision Tree

Off shore drilling operations will leave rock cuttings from the drilling process on the ocean floor. These cuttings are composed partially of petroleum bearing rock. The pile of cuttings left after the drilling is complete are called shell mounds because sea organisms with shells (among other animals) take up residence on the cuttings. Off the coast of central California an oil company had decommissioned the drilling platforms and wanted to be recused of further liability associated with the shell mounds they created. The question that the local citizens asked was "Should the shell mounds be removed, or should they stay in place in perpetuity? And if they stay in place, what is the risk given that the area is prone to earthquakes?"

This is a relatively complex problem that can be broken down into a decision tree for finding the optimal solution. Here the consequences were quantified in terms of surface area of the shell mound exposed, which is the percent area of a shell mound where the petroleum bearing rock is re-exposed due to seismic related deformations. This information was then combined into a larger decision tree that incorporated costs associated with sea life, fishing operations, beach tourism, and long term resident health.

A Decision Tree provided the framework and guidance for getting a first-order estimate of the risk due to earthquakes. The conditional probability of exposed surface area of the shell mound for a particular failure mode was calculated given the annual probability of exceedance of a large damaging earthquake. Based on offshore sampling and lab data, engineering calculations were combined with expert consensus to provide a relative assessment.

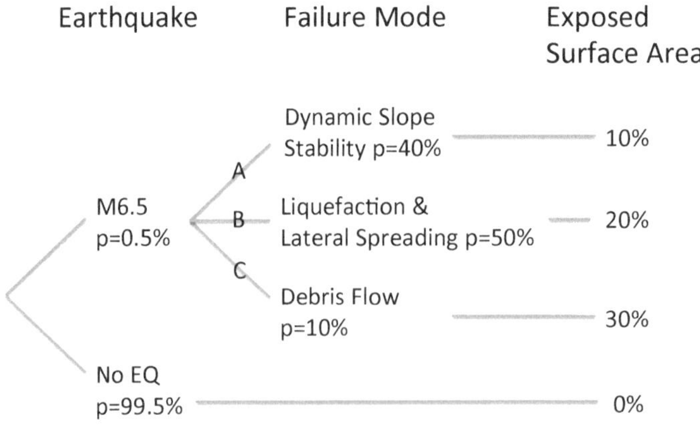

The large damaging earthquake (M6.5) is the maximum expected event for that area, and the annual probability of exceedance was calculated using a logic tree common in PSHA (probabilistic seismic hazard analysis; McGuire, 2004). The risk associated with each failure mode is:

$$R(A) = 0.005(0.40)0.10 = 0.00020$$
$$R(B) = 0.005(0.50)0.20 = 0.00050$$
$$R(C) = 0.005(0.10)0.50 = 0.00025$$

And the total annual risk due to an earthquake is the sum of the risk from each failure mode:

$$R(EQ) = 0.0002 + 0.0005 + 0.00025 = 0.00095$$

Compare this to 100% exposure in the year of removal if the shell mound were to be removed.

Armed with this rough analysis and subsequent details provided by a team of economists, the citizens made a decision to leave the shell mound in place and absolve the oil company of any future liability. The risk was effectively communicated to the stake holders and a rational decision was made that considered all the available information.

Example: Levee Failure Event Tree

The following event tree describes flooding potential associated with levee performance. Here we map the potential of poor planning that results in levee overtopping, and poor design that results in levee failure.

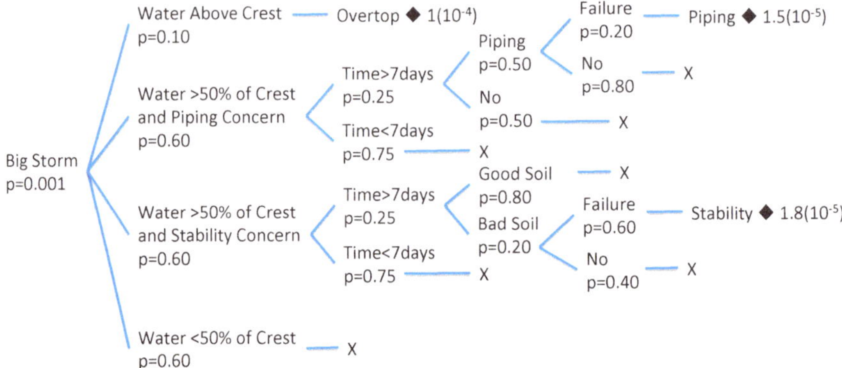

The event tree shows the relationship between the possible failure modes for this levee system. A branch of the tree that ends in an **X** denotes no failure. A branch of a tree that ends in failure has a ◆ and the associated probability.

In this example, the probability of a big storm, usually reported as an annual probability, is 0.1%, which is probably based on historical weather and stream flow data for the region. The overtopping here has the highest probability of occurrence when compared to piping and stability failure. This means that the peak water level (i.e., loading) for which the levee was designed was not adequately characterized, as opposed to piping and stability failures where the engineering of the levee (resistance) is faulty.

Example: Seismic Hazard Logic Tree

Logic Trees are used frequently for the problem of assessing seismic hazard. The annual probability of exceeding some ground shaking level is calculated as a multifold integral of conditional probabilities of earthquake rupture, magnitude, and distance. There are many competing models for calculating the seismic hazard that each contain some epistemic uncertainty. To defeat the epistemic uncertainty these competing models are used and weighted according to accuracy, confidence, or some other scheme. The following Logic Tree shows one path of a seismic hazard calculation that would be performed using Monte Carlo simulations to generate many realizations. The fault branch can include many faults in a region that could rupture and the probability associated with the fault represents the confidence in how viable the fault is in producing a rupture (this can include alternate fault geometry, fault segmentation, or other fault characteristics). The recurrence model branch includes two common relationships weighted according to how well they describe that particular fault. The slip rate and maximum magnitude branches are represented by a single model each with some defined uncertainty. And the ground motion prediction equation branch includes four equally viable models that are all included to defeat epistemic uncertainty in the individual models and produce a robust mean or median result.

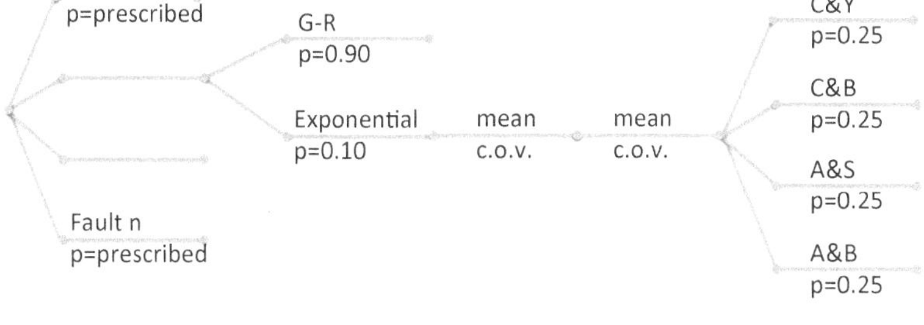

Example: Bridge Failure Fault Tree

In this example we are trying to quantify the probability of failure of an event that comprised of many subevents. The fault tree allows us to intuitively diagram the different subevents to then produce a statement of the total probability of the event. The different subevents, here different failure modes of a bridge, are shown connected by symbols that represent **or** (indicating union) or **and** (indicating intersection). As shown in the following Fault Tree, bridge collapse can be caused by a failed support **or** a failed pile **or** an overstressed girder. These events are then broken down into further subevents. We can then replace the description of the event and subevents with event numbers to aid us in writing the total probability statement. The order of the numbering is unimportant, it is just a short hand way of keeping track of the subevents.

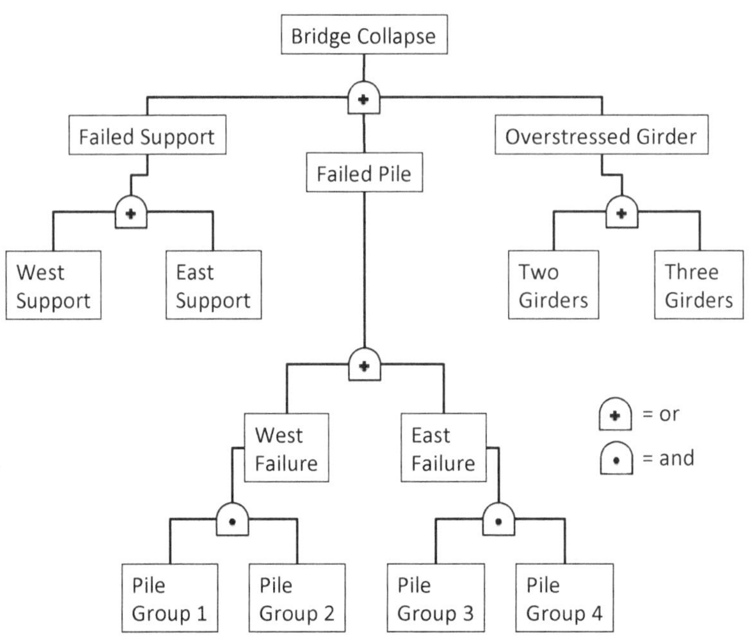
Bridge Collapse
Failed Support
Failed Pile
Overstressed Girder
West Support
East Support
Two Girders
Three Girders
West Failure
East Failure
= or
= and
Pile Group 1
Pile Group 2
Pile Group 3
Pile Group 4

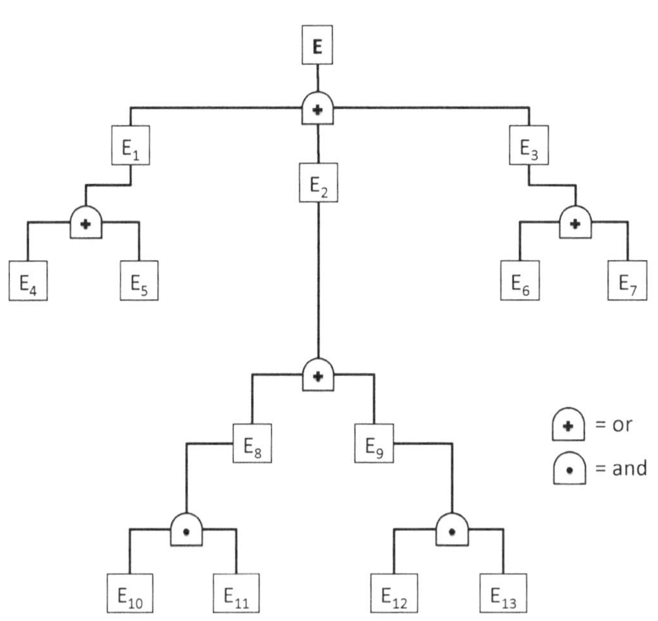
E
E_1
E_2
E_3
E_4
E_5
E_6
E_7
E_8
E_9
= or
= and
E_{10}
E_{11}
E_{12}
E_{13}

Using the numbered event tree we can write a probability statement describing the system as shown below. At the first level we have:

$$P(E) = P(E_1 \cup E_2 \cup E_3)$$

We take that to the second level of subevents:

$$P(E) = P((E_4 \cup E_5) \cup (E_8 \cup E_9) \cup (E_6 \cup E_7))$$

And the final level of subevents for this event tree are:

$$P(E) = P((E_4 \cup E_5) \cup ((E_{10} E_{11}) \cup (E_{12} E_{13})) \cup (E_6 \cup E_7))$$

The event tree allowed us to systematically piece together the complex probability statement of bridge collapse including several different failure modes. Component reliability analysis would inform us as to the probability of failure for each subevent which can then be plugged into this complex probability statement to arrive at the total probability of failure of the bridge, P(E).

Chapter Summary

- Complex or multi-component systems can often be intuitively mapped using different trees.
- Trees can allow for a broader perspective on a risk or probability of failure problem, thereby providing a framework for the component or system reliability analysis.
- Event Trees and Decision Trees map the sequence of events that lead up to failure. Event Trees quantify the probability of failure for a system and Decision Trees include consequences to provide risk for a system.
- Logic Trees are useful when multiple models can be used to solve a problem. The tree provides a weighting framework for including all models, thereby minimizing epistemic uncertainty in each individual model.
- Fault Trees can allow for a much clearer understanding of multiple failure modes in a system, and can provide a means of determining a total probability statement for the system.

9 RELIABILITY-BASED CODES (LRFD)

There are many design codes used in engineering design today. Originally most codes used some form of an allowable or working stress design (ASD/WSD) approach.

$$\frac{R}{FS} \geq \sum Q_i \qquad \text{equation 9.1}$$

R is the resistance, FS is the factor of safety, and Q_i are the loads (e.g., dead, live, extreme...). The factor of safety represents the lumped uncertainty from load and resistance and the desire for overdesign and safety in one factor. This approach has been used to check and ultimate limit state (i.e., failure, collapse, …) and/or a service limit state (i.e., deformation, deflection, …).

Within the last few decades design codes have been switching to a load and resistance factor design (LRFD) approach. The basis of LRFD is reliability, but the reason for the switch is because using a reliability based approach provides a more accurate answer which often results in cost savings for the project.

$$\varphi R \geq \sum \alpha_i Q_i \qquad \text{equation 9.2}$$

R and Qi are the same resistance and loads as described above, but instead of one lumped factor of safety we have φ the resistance factor and α_i the load factors. The resistance is factored down and the loads factored up to achieve a desired design level. The design level in LRFD is based on the reliability index thereby pinning the reliability on the number of standard deviations away from failure the design should be. This provides a more rational relationship between load and resistance by accounting for the respective uncertainties in each.

Reliability Formulation

There are two methods for determining the load and resistance factors for LRFD design:

1) Optimizing the LRFD equation using empirical data and reliability methods (FOSM and/or FORM). A target reliability index is used, often in the range of 2.0 to 3.0.
2) Calibrating the LRFD equation to previous ASD/WSD design practice using codified FS.

The first method is preferred, but there is often insufficient data to carry this out. The second method is comforting because it relies on past practice but may not achieve the goal of a more accurate answer and cost savings that the first method can. Examples of each are shown below.

Calibrated LRFD (Method 2)

We will start with method 2 where we are calibrating load and resistance factors from past codified factors of safety. If we combine equations 9.1 and 9.2 and solve for the resistance factor:

$$\varphi \geq \frac{\sum \alpha_i Q_i}{FS \sum Q_i} \quad \text{equation 9.3}$$

If only dead and live load are considered:

$$\varphi = \frac{\alpha_D Q_D + \alpha_L Q_L}{FS(Q_D + Q_L)} \quad \text{equation 9.4}$$

By dividing through by Q_L we get an equation in terms of the load ratio:

$$\varphi = \frac{\alpha_D \frac{Q_D}{Q_L} + \alpha_L}{FS(\frac{Q_D}{Q_L} + 1)} \quad \text{equation 9.5}$$

If we fix the load factors (that means fixing the uncertainty from dead and live load) then we can back calculate the resistance factor for a given factor of safety. Typical load factors from AASHTO (American Association of State Highway Transportation Officials who have implemented LRFD in highway bridge design) are α_D=1.25 and α_L=1.75. Fixing the load factors means that the remaining uncertainty in the factor of safety is attributed to the resistance;

we are in essence partitioning the lumped uncertainty in the factor of safety into load and resistance sides of the equation.

Table 9.1: Factor of Safety calibrated resistance factors (φ) for different dead (Q_D) versus live load (Q_L) ratios.

Calibrated φ Factors from Past FS			
FS	**Q_D/Q_L=1**	**Q_D/Q_L =2**	**Q_D/Q_L =3**
1.5	1.00	0.94	0.92
2.0	0.75	0.71	0.69
2.5	0.60	0.57	0.55
3.0	0.50	0.47	0.46

The loads are to be factored up by 125% and 175% respectively. If our ratio of dead to live load is 1 and the ASD/WSD design traditionally used a FS=3.0 then we will factor our resistance down by 50% to proceed with the design using LRFD (equation 9.2).

Optimized LRFD (Method 1)

If there is sufficient data to perform statistics then the load and resistance factors can be evaluated directly from the data, and not back calculated from prior factors of safety. This is the preferred method but requires sufficient data and careful study of the particular engineering problem.

For this method we will use a shallow foundation design problem (borrowed from Baecher and Christian, 2003) and FOSM to illustrate the application. The LRFD equation 9.2 (rewritten here):

$$\varphi R \geq \sum \alpha_i Q_i$$

We treat R and Q as random variables and in this problem will assume we are dealing with a single load to simplify the example:

$$\varphi \mu_R \geq \alpha \mu_Q \qquad \text{equation 9.7}$$

And if both R and Q are Gaussian (to allow for an exact solution) we know from Chapter 6 on Component Reliability that the margin of safety and reliability index are:

$$M = R - Q$$

$$\beta = \frac{\mu_M}{\sigma_M} = \frac{\mu_R - \mu_Q}{\sqrt{\sigma_R^2 + \sigma_Q^2 - 2\rho_{RQ}\sigma_R\sigma_Q}}$$

Solving for μ_R and substituting into the single load LRFD equation 9.7, we arrive at an expression for the resistance factor:

$$\varphi = \frac{\alpha\mu_Q}{\mu_R} = \frac{\alpha\mu_Q}{\beta\sqrt{\sigma_R^2 + \sigma_Q^2 - 2\rho_{RQ}\sigma_R\sigma_Q} + \mu_Q} \qquad \text{equation 9.8}$$

The reliability index is set at a target value, usually 2 or 3 depending on the number of standard deviations away from failure that is deemed acceptable for the particular engineering design. A target reliability index, like a factor of safety, is determined via consensus usually through building code panels or other expert consensus forums.

That leaves determining the moments of the load and resistance in addition to the load factor in order to solve for the resistance factor. Note that equation 9.8 applies to any similar engineering design situation where we are dealing with a single load and we have normally distributed load and resistance.

In this example we are analyzing the bearing capacity of a shallow foundation footing. For this problem R is the soil resistance or in geotechnical terms the ultimate bearing capacity (q_{ult}) as calculated using Terzaghi's method, Q is the single dead load due to the foundation often called the bearing pressure (q), and the load factor found in AASHTO foundation design recommendations, as discussed previously.

Terzaghi's method for calculating the soil resistance uses the following equation:

$$R = q_{ult} = cN_c + \sigma'_{ZD}N_q + \frac{1}{2}\gamma BN_\gamma$$

We will assume it is a strip footing ($B = 1m$) with its base located at the soil surface ($\sigma'_{ZD} = 0$) resting on a sandy soil ($c = 0$) with a unit weight (γ) of 18.5 kN/m^3 and a friction angle (ϕ) of 35 degrees. For readers unfamiliar with bearing capacity analysis the details of the equation are unimportant for this discussion, only the fact that the equation is a function of a random variable representing resistance is important. The bearing capacity equation reduces when we input the assumptions to:

$$R = q_{ult} = \frac{1}{2}\gamma B N_\gamma$$

The above equation is a function of the unit weight (γ), the footing dimension (B), and an empirical factor (N_γ). N_γ is itself a function of the friction angle (ϕ), the solution geometry, and the physical assumptions behind Terzaghi's method. If we treat N_γ as a random variable we can propagate the uncertainty it contains through the bearing capacity equation. N_γ is usually determined using linear regression which is a "best fit" to data that has a certain amount of scatter as shown below.

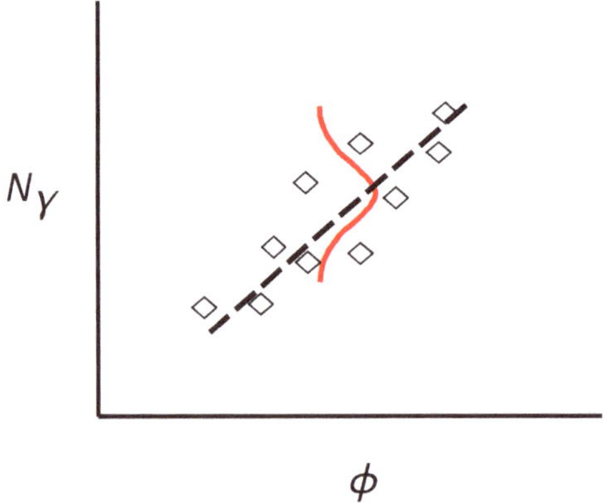

Figure 9.1 Schematic showing linear regression results of friction angle versus the empirical bearing capacity factor. The scatter or uncertainty in the relationship is evident and a hypothetical distribution is overlaid.

Using results from Terzaghi's (1943) original analysis for $\phi = 35°$ the statistics of the empirical factor are $\mu_{N_\gamma} = 82$ and $\sigma_{N_\gamma} = 17$. Using FOSM to estimate how this uncertainty propagates to q_{ult}:

$$\mu_{q_{ult}} \approx \frac{1}{2}\gamma B \mu_{N_\gamma} = 760kPa$$

$$\sigma_{q_{ult}} \approx \sqrt{\sigma_{N_\gamma}^2 \left(\frac{1}{2}\gamma B\right)^2} = 157kPa$$

$$\delta_{q_{ult}} \approx 21\%$$

We now have the resistance side of the problem defined by its first and second moments. For the load side of the problem the bearing pressure is usually calculated as a function of the column load (P), area (A), self weight (W) of the footing, and any uplift pore pressure (u) caused by a water table above the footing base:

$$q = \frac{P + W}{A} - u$$

For this example we will assume that the bearing pressure is roughly 250kPa per linear meter of the strip footing. To quantify the uncertainty of this load we look up a coefficient of variation reported in the structural codes, which gives $\delta_q \approx 15\%$. The codes also provide the load factor, and as mentioned we are just analyzing for a dead load $\alpha_D = 1.25$. Typical shallow foundation design would have a target reliability index of $\beta = 3.0$.

Neglecting any correlation between load and resistance we can calculate the resistance factor using equation 9.8:

$$\varphi = \frac{1.25(250)}{3.0\sqrt{157^2 + 37.5^2} + 250} = \frac{312.5}{734.3} = 0.43$$

This resistance factor is a function of the uncertainty, and here the uncertainty is from the method used to determine the bearing capacity. Table 9.2 shows a comparison of this calculated value to typical values (Barker et al., 1991; Withiam et al., 2001).

A semi-empirical method uses field test results directly for the resistance calculation (e.g., SPT blow counts). A rational method uses engineering properties that were converted from field or lab test results for the resistance calculation (e.g., friction angle from SPT blow counts).

For shallow foundation design; the soil type, method of analysis, and type of test used to measure the soil properties all affect the amount of uncertainty in the resultant, which is reflected in the value of the resistance factor. In LRFD design, the higher the resistance factor, the higher the confidence in the answer because there is less uncertainty in the variables and/or the resultant.

Shown in Table 9.2 are the factors of safety typically used for ASD design and the true or LRFD-equivalent factors of safety that include the uncertainty in the analysis. In most cases, shallow foundations are overdesigned using ASD because the uncertainty is not properly quantified. This is the impetus for quantifying the uncertainty and the basis of LRFD design: to better define the threshold of failure and properly design against it. Overdesign is often costly, both in time and materials, and by performing a more accurate analysis by including the uncertainty a better more cost-effective design can often be achieved.

Table 9.2: Typical resistance factors for shallow foundation bearing capacity analysis.

Soil Type	Method +Test	Resistance Factor ϕ	FS ASD	FS LRFD
Sand	Semi-empirical + SPT	0.45	3.0	3.2
	Semi-empirical + CPT	0.55	2.5	2.6
	Rational + SPT	0.35	2.5	4.1
	Rational + CPT	0.45	2.5	3.2
Clay	Semi-empirical + CPT	0.50	2.5	2.9
	Rational + lab	0.60	2.5	2.4
	Rational + field vane	0.60	2.5	2.4
	Rational + CPT	0.50	2.5	2.9

Chapter Summary

- LRFD is a design code framework that includes quantified uncertainty in the load and the resistance.
- The basis for LRFD is reliability, the same methods that were presented in Chapter 6.
- Load and Resistance Factors can be derived from statistics of the problem (Method 1) or back calculated from a previous Factor of Saftey (Method 2).
- Using LRFD provides a more accurate answer to the design problem, which translates to a safer design that also often results in cost-savings.

REFERENCES

Ang, A. H-S., and Tang, W. H. (1975). Probability Concepts in Engineering Planning and Design: Volume I-Basic Principles. Wiley, New York.

Ang, A. H-S., and Tang, W. H. (1984). Probability Concepts in Engineering Planning and Design: Volume II-Decision, Risk, and Reliability. Published by the authors.

Ang, A. H-S. and Tang, W. H. (2007). Probability Concepts in Engineering: Emphasis on Applications to Civil and Environmental Engineering, 2nd Edition. Wiley, New York.

Baecher, G.B. and Christian, J.T. (2003). Reliability and Statistics in Geotechnical Engineering. Wiley. New Jersey.

Barker, R.M., J.M. Duncan, K.B. Rojiani, P.S.K. Ooi, C.K. Tan, and S.G. Kim (1991) "Load Factor Design Criteria for Highway Structure Foundations." Final Report, NCHRP Project 24-4, Virginia Polytechnic Institute and State University, Blacksburg, VA.

Benjamin, J.R., and Cornell, C.A. (1970). Probability, Statistics, and Decisions for Civil Engineers. McGraw-Hill, New York.

Bernstein, P.L. (1998) Against the Gods: The Remarkable Story of Risk. Wiley

Borri, A., Ceccotti, A., and Spinelli, P. (1983). "Statistical Analysis of the Influence of Timber Defects on the Static Behavior of Glue Laminated Beams." Proc. 4th Int. Conf. Applications of Statistics and Probability in Soil and Structural Engineering, Vol.1, June.

Der Kiureghian, A., Lin, H.-Z., Hwang, S.-J. (1987). "Second-order reliability approximations." Journal of Engineering Mechanics, ASCE, 113(8), 1208-1225.

Der Kiureghian, A. (2001). Structural Reliability Course Notes. UC Berkeley.

De Finetti, B (1972). Probability, Induction and Statistics: The Art of Guessing. Wiley, London.

De Finetti, B (1974). Theory of Probability: A Critical Introductory Treatment. Wiley. London.

Ditlevsen, O. (1979). Narrow reliability bounds for structural systems. J. Structural Mechanics, 7(4), 453-472.

Ellingwood, B., Galambos, T.V., MacGregor, J.G., and Cornell, C.A. (1980). "Development of a Probability-Based Load Criteriaon for the American National Standard A58." Special Pub. 577, National Bureau of Standards, Washington D.C., June.

Fredlund, D.G., and Dahlman, A.E. (1972). Statistical Geotechnical Properties of Glacial Lake Edmonton Sediments." Statistics and Probability in Civil Engineering, Oxford University Press, London.

Gonick, L., and Smith, W. (1993). The Cartoon Guide to Statistics. Harper Collins Publishers Inc., New York.

Hammitt, G.M. (1966). "Statistical Analysis of Data from a Comparative Laboratory Test Program Sponsored by ACIL." Misc. Paper 4-785, U.S. Army Engineering Waterways Experiment Station, Corps of Engineers.

Hasofer A.M., and Lind, N. (1974) "An Exact and Invariant First-Order Reliability Format." Journal of Engineering Mechanics, ASCE, 100(EM1), Februrary, 111-121.

Hollenback, J. C. (2013). Reliability of levee systems under seismic loading. Dissertation in partial fulfillment of Ph.D., U.C. Berkeley.

Hollenback, J. C., and Moss, R.E.S. (2011). "Bounding the probability of failure for levee systems." Proc. GeoRisk 2011, ASCE, Atlanta.

Holmes, R.M., and De Burger, J. (1988). Serial Murder. Sage Publications, Beverly Hills, CA.

Hunter, D. (1976). An upper bound for the probability of a union. J. Applied Probability, 13, 597-603.

Knuth, D. (1997). "Chapter 3 – Random Numbers". The Art of Computer Programming. Vol. 2: Seminumerical algorithms (3 ed.).

Kounias, E.G. (1968). Bound for the probability of a union, with applications. Amer. Math. Stat., 39(6), 2154-2158.

Kulhawy, F.H., and Mayne, P.W. (1990). Manual on Estimating Soil Properties for Foundation Design. EPRI EL-6800 Research Project 1493-6, Aug.

Laplace, P.S. (1812). "Théorie analytique des probabilities." Original tretise, Paris.

Laplace, P. S. (1814). "Essai philosophique sur les probabilities." Gauthier-Villars, Paris. Modern translated version; Laplace (1951). "A philosophical Essay on Probabilities." Dover, Toronto.

Lichtenstein, S., Fischhoff, B., and Phillips, L. (1982). "Calibration of probabilities: The state of the art to 1980." Judgement under Uncertainty: Heuristics and Biases. Kahneman, D, Slivc, P. and Tversky, A, eds., Cambridge Press, New York, 306-334.

Low, B. K., and Tang, W. H. (1997). Efficient Reliability Evaluation Using Spreasheet. Journal of Engineering Mechanics, 123(7), 749-752.

Low, B. K., and Tang, W. H. (2004). "Reliability analysis using object-oriented constrained optimization." Structural Safety, 26(1), 69-89.

Low, B. K. (2005). Reliability-based design applied to retaining walls. Geotechnique, 55(1), 63-75.

Lumb, P. (1966). "Variability of Natural Soils." Canadian Geotechnical Journal, Vol.3, May.

Marsaglia, George (2004). "Evaluating the Normal Distribution". Journal of Statistical Software 11(4).

McGrayne, S.B. (2011). The theory that would not die. How Bayes' rule cracked the enigma code, hunted down Russian submarines, and emerged triumphant from two centuries of controversy. Yale University Press.

McGuire, R.K. (2004). Seismic Hazard and Risk Analysis. Earthquake Engineering Research Institute (EERI) Monograph MNO-10.

Moss, R. E. S. (2008). "Quantifying Measurement Uncertainty of Thirty Meter Shear Wave Velocity (VS30)." Bulletin of Seismological Society of America, 98(3), 1399-1411.

Moss, R.E.S. (2009) Reduced Uncertainty of Ground Motion Prediction Equations through Bayesian Variance Analysis. Pacific Earthquake Engineering Research Center (PEER) Report#105, Appendix on Bayesian regression. http://peer.berkeley.edu/publications/peer_reports/reports_2009/reports_2009.html

Moss, R. E. S. (2011). "Reduced Sigma of Ground Motion Prediction Equations through Uncertainty Propagation." Bulletin of Seismological Society of America, 101(1).

Moss, R.E.S., and Germeraad, M. (2013). Lifelines Annex to the 2013 California State multi-Hazard Mitigation Plan. http://hazardmitigation.calema.ca.gov/plan/state_multi-hazard_mitigation_plan_shmp

Nielsen, D.R., Biggar, J.W., and Erh, K.T. (1974). Spatial Variability of Field-Measured Soil-Water Properties." Hilgardia, J. Agr. Sci. (CA Agr. Experiment Sta.), Vol.42, pp. 215-260, Nov.

Padilla, J.D., and Vanmarcke, E.H. (1974). Settlement of Structures on Shallow Foundations: A Probabilistic Analysis." Res. Rep. R74-9, MIT, Cambridge, Mass.

Ripley, B.D. (1987) Stochastic Simulation, 1987, Wiley & Sons, New York.

Rosenbluth, E. (1975). "Point estimates for probability moments." Proc., National Academy of Science, 72(10), 3812-3814.

Rosenbluth, E. and Esteva, L. (1972). "Reliability basis for some Mexican Codes." American Concrete Institute, Detroit.

Phoon, K-K, and Nadim, F. (2004). Modeling Non-Gaussian Random Vectors for FORM: State-of-the-Art Review. Keynote paper Int. Workshop on Risk Assessment in Site Characterization and Geotechnical Design, Bangalore, India, November, 55-85.

Schultze, E. (1972). "Frequency Distributions and Correlations of Soil Properties." Statistics and Probability in Civil Engineering, Oxford University Press, London.

Schultze, E. (1975). "The General Significance of Statistics for Civil Engineering." Proc. 2nd Int. Conf. Applications of Statistics and Probability in Soil and Structural Engineering, Germany, Sept.

Small, M., and Singer, J.D. (1982). Resort to Arms: International and Civil Wars 1816-1980. Sage Publications, Beverly Hills, CA.

Song, J., and Der Kiureghian, A. (2003). "Bounds on system reliability by linear programming." J. Engineering Mechanics, ASCE, 129(6): 627-636.

Terzaghi, K. (1943). Theoretical Soil Mechanics. John Wiley and Sons, New York.

U.S. Nuclear Regulatory Commission. (1975). "Reactor Safety Study: An Assessment of accident risks in the U.S. commercial nuclear power plants." NRC WASH-1400, Washington D.C.

Wilson, R., and Crouch, E.A. (1987). "Rish Assessment and Comparisons: An Introduction." Science, Vol. 236, no. 4799, pp. 267-270, April.

Withiam, J.L. , et al. (2001). "Load and resistance factor design (LRFD) for highway bridge substructures." Report No. FHWA HI-98-032, Federal Highway Administration, Washington, D. C

Zhang, Y., and Der Kiureghian, A. (1995). "Two improved algorithms for reliability analysis." Reliability of and Optimization of Structural Systems. Proc., 6th IFIP working conference on optimization of structural systems, 297-204.

APPENDIX A
STANDARD NORMAL PROBABILITY TABLES

There is no closed-form integral for the normal distribution. Because it is so frequently used approximations of the CDF are commonly found in most computational devices. The following two pages provide tabular results of the standard normal cumulative distribution function, N(0,1). In the tables x is the value of interest and $\Phi(x)$ is the integral of the standard normal PDF at that value. The x represents the number of standard deviations away from the mean the value is located, so a value of 1.00 is approximately the 84th percentile or 84% probability, the value 2.00 is approximately the 98th percentile or 98% probability, and so on.

The standard normal CDF can also be approximated using many equations that can be found in the literature. A simple function based on a Taylor series expansion is shown below (Marsaglia, 2004) with its accuracy a function of the number of terms used in the expansion:

$$\Phi(x) \cong \frac{1}{2} + \phi(x)\left(x + \frac{x^3}{3} + \frac{x^5}{3 \cdot 5} + \frac{x^7}{3 \cdot 5 \cdot 7} + \frac{x^9}{3 \cdot 5 \cdot 7 \cdot 9} + \cdots\right)$$

x	Φ(x)	x	Φ(x)	x	Φ(x)
0.00	0.5000	0.31	0.6217	0.62	0.7324
0.01	0.5040	0.32	0.6255	0.63	0.7357
0.02	0.5080	0.33	0.6293	0.64	0.7389
0.03	0.5120	0.34	0.6331	0.65	0.7422
0.04	0.5160	0.35	0.6368	0.66	0.7454
0.05	0.5199	0.36	0.6406	0.67	0.7486
0.06	0.5239	0.37	0.6443	0.68	0.7517
0.07	0.5279	0.38	0.6480	0.69	0.7549
0.08	0.5319	0.39	0.6517	0.70	0.7580
0.09	0.5359	0.40	0.6554	0.71	0.7611
0.10	0.5398	0.41	0.6591	0.72	0.7642
0.11	0.5438	0.42	0.6628	0.73	0.7673
0.12	0.5478	0.43	0.6664	0.74	0.7704
0.13	0.5517	0.44	0.6700	0.75	0.7734
0.14	0.5557	0.45	0.6736	0.76	0.7764
0.15	0.5596	0.46	0.6772	0.77	0.7794
0.16	0.5636	0.47	0.6808	0.78	0.7823
0.17	0.5675	0.48	0.6844	0.79	0.7852
0.18	0.5714	0.49	0.6879	0.80	0.7881
0.19	0.5753	0.50	0.6915	0.81	0.7910
0.20	0.5793	0.51	0.6950	0.82	0.7939
0.21	0.5832	0.52	0.6985	0.83	0.7967
0.22	0.5871	0.53	0.7019	0.84	0.7995
0.23	0.5910	0.54	0.7054	0.85	0.8023
0.24	0.5948	0.55	0.7088	0.86	0.8051
0.25	0.5987	0.56	0.7123	0.87	0.8078
0.26	0.6026	0.57	0.7157	0.88	0.8106
0.27	0.6064	0.58	0.7190	0.89	0.8133
0.28	0.6103	0.59	0.7224	0.90	0.8159
0.29	0.6141	0.60	0.7257	0.91	0.8186
0.30	0.6179	0.61	0.7291	0.92	0.8212

x	Φ(x)
0.93	0.8238
0.94	0.8264
0.95	0.8289
0.96	0.8315
0.97	0.8340
0.98	0.8365
0.99	0.8389
1.00	0.8413
1.01	0.8438
1.02	0.8461
1.03	0.8485
1.04	0.8508
1.05	0.8531
1.06	0.8554
1.07	0.8577
1.08	0.8599
1.09	0.8621
1.10	0.8643
1.11	0.8665
1.12	0.8686
1.13	0.8708
1.14	0.8729
1.15	0.8749
1.16	0.8770
1.17	0.8790
1.18	0.8810
1.19	0.8830
1.20	0.8849
1.21	0.8869
1.22	0.8888
1.23	0.8907

x	Φ(x)
1.24	0.8925
1.25	0.8944
1.26	0.8962
1.27	0.8980
1.28	0.8997
1.29	0.9015
1.30	0.9032
1.31	0.9049
1.32	0.9066
1.33	0.9082
1.34	0.9099
1.35	0.9115
1.36	0.9131
1.37	0.9147
1.38	0.9162
1.39	0.9177
1.40	0.9192
1.41	0.9207
1.42	0.9222
1.43	0.9236
1.44	0.9251
1.45	0.9265
1.46	0.9279
1.47	0.9292
1.48	0.9306
1.49	0.9319
1.50	0.9332
1.51	0.9345
1.52	0.9357
1.53	0.9370
1.54	0.9382

x	Φ(x)
1.55	0.9394
1.56	0.9406
1.57	0.9418
1.58	0.9429
1.59	0.9441
1.60	0.9452
1.61	0.9463
1.62	0.9474
1.63	0.9484
1.64	0.9495
1.65	0.9505
1.66	0.9515
1.67	0.9525
1.68	0.9535
1.69	0.9545
1.70	0.9554
1.71	0.9564
1.72	0.9573
1.73	0.9582
1.74	0.9591
1.75	0.9599
1.76	0.9608
1.77	0.9616
1.78	0.9625
1.79	0.9633
1.80	0.9641
1.81	0.9649
1.82	0.9656
1.83	0.9664
1.84	0.9671
1.85	0.9678

x	Φ(x)
1.86	0.9686
1.87	0.9693
1.88	0.9699
1.89	0.9706
1.90	0.9713
1.91	0.9719
1.92	0.9726
1.93	0.9732
1.94	0.9738
1.95	0.9744
1.96	0.9750
1.97	0.9756
1.98	0.9761
1.99	0.9767
2.00	0.9772
2.01	0.9778
2.02	0.9783
2.03	0.9788
2.04	0.9793
2.05	0.9798
2.06	0.9803
2.07	0.9808
2.08	0.9812
2.09	0.9817
2.10	0.9821
2.11	0.9826
2.12	0.9830
2.13	0.9834
2.14	0.9838
2.15	0.9842
2.16	0.9846

x	Φ(x)
2.17	0.9850
2.18	0.9854
2.19	0.9857
2.20	0.9861
2.21	0.9864
2.22	0.9868
2.23	0.9871
2.24	0.9875
2.25	0.9878
2.26	0.9881
2.27	0.9884
2.28	0.9887
2.29	0.9890
2.30	0.9893
2.31	0.9896
2.32	0.9898
2.33	0.9901
2.34	0.9904
2.35	0.9906
2.36	0.9909
2.37	0.9911
2.38	0.9913
2.39	0.9916
2.40	0.9918
2.41	0.9920
2.42	0.9922
2.43	0.9925
2.44	0.9927
2.45	0.9929
2.46	0.9931
2.47	0.9932

x	Φ(x)
2.48	0.9934
2.49	0.9936
2.50	0.9938
2.51	0.9940
2.52	0.9941
2.53	0.9943
2.54	0.9945
2.55	0.9946
2.56	0.9948
2.57	0.9949
2.58	0.9951
2.59	0.9952
2.60	0.9953
2.61	0.9955
2.62	0.9956
2.63	0.9957
2.64	0.9959
2.65	0.9960
2.66	0.9961
2.67	0.9962
2.68	0.9963
2.69	0.9964
2.70	0.9965
2.71	0.9966
2.72	0.9967
2.73	0.9968
2.74	0.9969
2.75	0.9970
2.76	0.9971
2.77	0.9972
2.78	0.9973

x	Φ(x)
2.79	0.9974
2.80	0.9974
2.81	0.9975
2.82	0.9976
2.83	0.9977
2.84	0.9977
2.85	0.9978
2.86	0.9979
2.87	0.9979
2.88	0.9980
2.89	0.9981
2.90	0.9981
2.91	0.9982
2.92	0.9982
2.93	0.9983
2.94	0.9984
2.95	0.9984
2.96	0.9985
2.97	0.9985
2.98	0.9986
2.99	0.9986
3.00	0.9987
...	...
3.50	0.99977
...	...
4.00	0.99997

APPENDIX B
FUNCTION MEAN AND VARIANCE PROOF

The following derivation is for the moments (mean and variance) of a function of random variables.

Take a function of two continuous random variables:

$$Z = g(X, Y)$$

The CDF of Z is then the integral of the joint distribution of X and Y considering the functional relationship of the variables:

$$F(z) = \iint_{g(x,y) \le z} f(x, y) dxdy = \int_{-\infty}^{\infty} \int_{-\infty}^{g^{-1}(z,y)} f(x, y) dxdy$$

To include the functional relationship we solve the function for x, $g^{-1} = g^{-1}(z, y)$, and change the variable of integration from x to z:

$$F(z) = \int_{-\infty}^{\infty} \int_{-\infty}^{z} f(g^{-1}, y) \left| \frac{\partial g^{-1}}{\partial z} \right| dzdy$$

To find the PDF of Z we take the derivative of the CDF with respect to z:

$$f(z) = \int_{-\infty}^{\infty} f(g^{-1}, y) \left| \frac{\partial g^{-1}}{\partial z} \right| dy$$

Alternately we could solve the functional form for for y instead of x, $g^{-1} = g^{-1}(x, z)$, and find:

$$f(z) = \int_{-\infty}^{\infty} f(x, g^{-1}) \left| \frac{\partial g^{-1}}{\partial z} \right| dx$$

The above results are applicable to any function with random variables of any distribution. Solving for this PDF (i.e., propagating the error from the independent variables to the dependent variable) can be simple in certain situations as pointed out in Chapter 5, or can be difficult when we have a functions that is difficult to find the inverse of and/or when we have random variables with distributions that are intractable.

In the case where our function is a sum, $Z = X + Y$, we find:

$$x = z - y$$

$$\frac{\partial g^{-1}}{\partial z} = \frac{\partial x}{\partial z} = 1$$

Resulting in the PDF:

$$f(z) = \int_{-\infty}^{\infty} f(z-y, y)\, dy$$

or alternatively:

$$f(z) = \int_{-\infty}^{\infty} f(x, z-x)\, dx$$

If X and Y are statistically independent then we can evaluate using their marginal distributions:

$$f(z) = \int_{-\infty}^{\infty} f(z-y) f(y)\, dy$$

or alternatively:

$$f(z) = \int_{-\infty}^{\infty} f(x) f(z-x)\, dx$$

Now if X and Y are Gaussian (i.e., Normally distributed) the PDF becomes:

$$f(z) = \frac{1}{2\pi\sigma_X\sigma_Y}\int_{-\infty}^{\infty} exp\left[-\frac{1}{2}\left(\frac{z-y-\mu_X}{\sigma_X}\right)^2 - \frac{1}{2}\left(\frac{x-\mu_Y}{\sigma_Y}\right)^2\right] dy$$

$$= \frac{1}{2\pi\sigma_X\sigma_Y} exp\left[-\frac{1}{2}\left\{\left(\frac{\mu_Y}{\sigma_Y}\right)^2 + \left(\frac{z-\mu_X}{\sigma_X}\right)^2\right\}\right] \int_{-\infty}^{\infty} exp\left[-\frac{1}{2}(uy^2 - 2vy)\right] dy$$

where $u = \frac{1}{\sigma_X^2} + \frac{1}{\sigma_Y^2}$ and $v = \frac{\mu_Y}{\sigma_Y^2} + \frac{z-\mu_X}{\sigma_X^2}$

substituting $w = y - \frac{v}{u}$ the integral becomes

$$\int_{-\infty}^{\infty} exp\left[-\frac{1}{2}(uy^2 - 2vy)\right] dy$$
$$= e^{v^2/2u} \int_{-\infty}^{\infty} exp\left(-\frac{1}{2}uw^2\right) dw = \sqrt{2\pi/u}\, exp\left(v^2/2u\right)$$

Reducing the algebra in the full distribution reveals:

$$f(z) = \frac{1}{\sqrt{2\pi(\sigma_X^2 + \sigma_Y^2)}} exp\left[-\frac{1}{2}\left(\frac{z - (\mu_X + \mu_Y)}{\sqrt{\sigma_X^2 + \sigma_Y^2}}\right)^2\right]$$

which is the definition of the joint Normal PDF with the moments:

$$\mu_Z = \mu_X + \mu_Y$$

$$\sigma_Z^2 = \sigma_X^2 + \sigma_Y^2$$

This can be solved for in a similar manner with correlated X and Y resulting in the moments:

$$\mu_Z = \mu_X + \mu_Y$$

$$\sigma_Z^2 = \sigma_X^2 + \sigma_Y^2 + 2\rho\sigma_X\sigma_Y$$

If the function contains constants or coefficients in front of statistically independent variables, $Z = aX + bY$, the moments are then:

$$\mu_Z = a\mu_X + b\mu_Y$$

$$\sigma_Z^2 = a^2\sigma_X^2 + b^2\sigma_Y^2$$

Similarly we can evaluate a difference function, $Z = X - Y$, which is the same but with a negative coefficient in front of the second variable:

$$\mu_Z = \mu_X - \mu_Y$$

$$\sigma_Z^2 = \sigma_X^2 + \sigma_Y^2$$

This can be expand to accommodate multiple random variables with their respective coefficients,

$$Z = \sum_{i=1}^{n} a_i X_i$$

$$\mu_Z = \sum_{i=1}^{n} a_i \mu_{X_i}$$

$$\sigma_Z^2 = \sum_{i=1}^{n} a_i^2 \sigma_{X_i}^2$$

The canonical form of these equations can be found in Chapter 5 in the discussion on Notation Clarity.

For a product function, $Z = XY$, where X and Y are Lognormal with their respective moments λ and ξ, we can use the same derivation and find:

$$lnZ = lnX + lnY$$

$$\lambda_Z = \lambda_X + \lambda_Y$$

$\xi_Z^2 = \xi_X^2 + \xi_Y^2$ for statistically independent

$$\xi_Z^2 = \xi_X^2 + \xi_Y^2 + 2\rho\xi_X\xi_Y \text{ with correlation}$$

And similarly for a quotient function, $Z = X/Y$, where X and Y are Lognormal with their respective moments λ and ξ:

$$lnZ = lnX - lnY$$

$$\lambda_Z = \lambda_X - \lambda_Y$$

$\xi_Z^2 = \xi_X^2 + \xi_Y^2$ for statistically independent

$$\xi_Z^2 = \xi_X^2 + \xi_Y^2 - 2\rho\xi_X\xi_Y \text{ with correlation}$$

The also can be easily expanded to multiple Lognormal random variables with their respective coefficients.

APPENDIX C
FORM HLRF ALGORITHM

The following solution of the cut slope problem using FORM (first order reliability method) demonstrates the "improved" HLRF (Hasofer Lind and Rackwitz Fiessler) algorithm as presented in Zhang and Der Kiureghian (1995). MathCad was used to demonstrate as clearly as possible the matrix manipulations and how the algorithm works. The correlated and uncorrelated examples are shown to allow comparison with other solution methods shown in Chapter 6.

The solution uses matrix mathematics for efficient calculations. The matrices often used in this or similar solutions are shown below in reliability notation. These matrices, of course, can be expanded to any number of variables when doing multivariate limit state analysis.

Mean vector $$M = \begin{bmatrix} \mu_R \\ \mu_Q \end{bmatrix}$$

Covariance matrix $$\Sigma = \begin{bmatrix} \sigma_R^2 & \rho_{RQ}\sigma_R\sigma_Q \\ \rho_{RQ}\sigma_R\sigma_Q & \sigma_Q^2 \end{bmatrix}$$

Standard Deviation matrix $$D = \begin{bmatrix} \sigma_R & 0 \\ 0 & \sigma_Q \end{bmatrix}$$

Correlation matrix $$R = \begin{bmatrix} 1 & \rho_{RQ} \\ \rho_{RQ} & 1 \end{bmatrix}$$

The following calculations show the application of the "improved HLRF" algorithm (Zhang and Der Kiureghian, 1995) for FORM analysis. Shown is the vertical cut in soil example to demonstrate the steps involved. This algorithm can be applied to any reliability problem by inserting the function, g(x), the transformation of the variables to standard normal space, and the gradient of g(x).

The equation is $gx := c - \frac{H}{4}\cdot\gamma$ cohesion and gamma are correlated normals (jointly normal r.v.'s) with a correlation coefficient of 0.5.

$R := \begin{pmatrix} 1 & 0.5 \\ 0.5 & 1 \end{pmatrix}$ $D := \begin{pmatrix} 30 & 0 \\ 0 & 2 \end{pmatrix}$ R is the correlation matrix, D is the standard deviation matrix.

$L := \text{cholesky}(R)$ $L = \begin{pmatrix} 1 & 0 \\ 0.5 & 0.866 \end{pmatrix}$ We find the L matrix using Cholesky decomposition.

$M := \begin{pmatrix} 100 \\ 20 \end{pmatrix}$ $X := M$ $H := 10$ M is the vector of means. We set the initial trial of this iterative solution X equal to the mean values. H is a constant in this analysis.

Transformation

The variables are transformed into standard normal space, from x to u, using the procedure below.

$u := L^{-1}\cdot D^{-1}\cdot(X - M)$ $gx := M_0 - \frac{H}{4}\cdot M_1$

$\Delta gx := \begin{pmatrix} 1 \\ -\frac{H}{4} \end{pmatrix}$ The gradian vector has the partial derivatives of gx.

HLRF Algorithm

We solve for the Jacobian matrix as shown and follow the steps of the HLRF algorithm.

$Gu := gx$ $Jux := L^{-1}D^{-1}$ $Jxu := Jux^{-1}$

$\Delta Gu := Jxu^T\cdot\Delta gx$ $\Delta Gu = \begin{pmatrix} 27.5 \\ -4.33 \end{pmatrix}$

$\alpha := -\frac{\Delta Gu}{|\Delta Gu|}$ $\alpha = \begin{pmatrix} -0.988 \\ 0.156 \end{pmatrix}$

Direction Vector

$$d := \left(\frac{gx}{|\Delta Gu|} + \alpha \cdot u\right) \cdot \alpha - u \qquad d = \begin{pmatrix} -1.774 \\ 0.279 \end{pmatrix}$$

$step := 1$

New Point

$$unew := u + step \cdot d \qquad unew = \begin{pmatrix} -1.774 \\ 0.279 \end{pmatrix}$$

Check

$$\frac{|u|}{|\Delta Gu|} = 0 \qquad c := 1$$

$$m1 := 0.5 \cdot (|u|)^2 + c \cdot |gx| \qquad m1 = 50$$

evaluate gx for new point

$$m2 := 0.5 \cdot unew^2 + c \cdot gx \qquad m2 = \begin{pmatrix} 51.574 \\ 50.039 \end{pmatrix}$$

if m2<m1 then converging

The results in standard norm and original space are:

$$unew = \begin{pmatrix} -1.774 \\ 0.279 \end{pmatrix} \qquad \alpha = \begin{pmatrix} -0.988 \\ 0.156 \end{pmatrix}$$

$$\beta := \alpha^T \cdot unew \qquad \beta = 1.796$$

$$pf1 := pnorm(-|\beta|, 0, 1) \qquad pf1 = 0.0362$$

Solving the same vertical cut in soil problem but this time the r.v.'s are uncorrlated.

insert gx, transformations, Jacobian, and gradient of gx.

$$R := \begin{pmatrix} 1 & 0 \\ 0 & 1 \end{pmatrix} \qquad D := \begin{pmatrix} 30 & 0 \\ 0 & 2 \end{pmatrix}$$

$$L := \text{cholesky}(R) \qquad L = \begin{pmatrix} 1 & 0 \\ 0 & 1 \end{pmatrix}$$

$$M := \begin{pmatrix} 100 \\ 20 \end{pmatrix} \qquad X := M \qquad H := 10$$

Transformation

$$u := L^{-1} \cdot D^{-1} \cdot (X - M)$$

$$\Delta gx := \begin{pmatrix} 1 \\ -\frac{H}{4} \end{pmatrix} \qquad gx := M_0 - \frac{H}{4} \cdot M_1$$

HLRF Algorithm

$$Gu := gx \qquad Jux := L^{-1} D^{-1} \qquad Jxu := Jux^{-1}$$

$$\Delta Gu := Jxu^T \cdot \Delta gx \qquad \Delta Gu = \begin{pmatrix} 30 \\ -5 \end{pmatrix}$$

$$\alpha := -\frac{\Delta Gu}{|\Delta Gu|} \qquad \alpha = \begin{pmatrix} -0.986 \\ 0.164 \end{pmatrix}$$

Direction Vector

$$d := \left(\frac{gx}{|\Delta Gu|} + \alpha \cdot u\right)\cdot\alpha - u \qquad d = \begin{pmatrix} -1.622 \\ 0.27 \end{pmatrix}$$

$step := 1$

New Point

$$unew := u + step \cdot d \qquad unew = \begin{pmatrix} -1.622 \\ 0.27 \end{pmatrix}$$

Check

$$\frac{|u|}{|\Delta Gu|} = 0 \qquad c := 1$$

$$m1 := 0.5 \cdot (|u|)^2 + c \cdot |gx| \qquad m1 = 50$$

evaluate gx for new point

$$m2 := 0.5 \cdot unew^2 + c \cdot gx \qquad m2 = \begin{pmatrix} 51.315 \\ 50.037 \end{pmatrix}$$

if m2<m1 then converging

The results in standard norm and original space are:

$$unew = \begin{pmatrix} -1.622 \\ 0.27 \end{pmatrix} \qquad \alpha = \begin{pmatrix} -0.986 \\ 0.164 \end{pmatrix}$$

$$\beta := \alpha^T \cdot unew \qquad \beta = 1.644$$

$$pf1 := pnorm(-|\beta|, 0, 1) \qquad pf1 = 0.0501$$

INDEX

www.ingramcontent.com/pod-product-compliance
Lightning Source LLC
LaVergne TN
LVHW052251100826
845147LV00001B/13

* 9 7 8 0 9 8 9 8 8 9 6 0 5 *